STORAGE AND HYBRIDIZATION OF NUCLEAR ENERGY

STORAGE AND HYBRIDIZATION OF NUCLEAR ENERGY

Techno-economic Integration of Renewable and Nuclear Energy

Edited by

HITESH BINDRA

SHRIPAD REVANKAR

ELSEVIER

ACADEMIC PRESS

An imprint of Elsevier

Academic Press is an imprint of Elsevier
125 London Wall, London EC2Y 5AS, United Kingdom
525 B Street, Suite 1650, San Diego, CA 92101, United States
50 Hampshire Street, 5th Floor, Cambridge, MA 02139, United States
The Boulevard, Langford Lane, Kidlington, Oxford OX5 1GB, United Kingdom

Library of Congress Cataloging-in-Publication Data
A catalog record for this book is available from the Library of Congress

British Library Cataloguing-in-Publication Data
A catalogue record for this book is available from the British Library

ISBN 978-0-12-813975-2

Publisher: Jonathan Simpson
Acquisition Editor: Maria Convey
Editorial Project Manager: Peter Jardim
Production Project Manager: Surya Narayanan Jayachandran
Cover Designer: Victoria Pearson

Typeset by SPi Global, India

CONTENTS

CONTRIBUTORS

Hitesh Bindra
Department of Mechanical and Nuclear Engineering, Kansas State University, Manhattan, KS, United States

Shannon M. Bragg-Sitton
Systems Integration Department, Nuclear Systems Design and Analysis, Idaho National Laboratory, Idaho falls, ID, United States

Haisheng Chen
Institute of Engineering Thermophysics, Chinese Academy of Sciences, Beijing, China

Justin L. Coleman
Fusion Hydrogen and Measurement Science Department, Idaho National Laboratory, Idaho falls, ID, United States

Lin Cong
Birmingham Centre for Energy Storage, School of Chemical Engineering, University of Birmingham, Birmingham, United Kingdom

Yulong Ding
Birmingham Centre for Energy Storage, School of Chemical Engineering, University of Birmingham, Birmingham, United Kingdom; School of Energy and Environmental Engineering, University of Science and Technology Beijing, Beijing, China

Eric J. Dufek
Energy Storage and Advanced Vehicles Department, Idaho National Laboratory, Idaho falls, ID, United States

Abhinav Gairola
Department of Mechanical and Nuclear Engineering, Kansas State University, Manhattan, KS, United States

Kathryn D. Huff
University of Illinois at Urbana-Champaign, Urbana, IL, United States

Samuel C. Johnson
The University of Texas at Austin, Austin, TX, United States

Yongliang Li
Birmingham Centre for Energy Storage, School of Chemical Engineering, University of Birmingham, Birmingham, United Kingdom

Xiaodong Peng
Birmingham Centre for Energy Storage, School of Chemical Engineering, University of Birmingham, Birmingham, United Kingdom

Shripad T. Revankar
School of Nuclear Engineering, Purdue University, West Lafayette, IN, United States

Joshua D. Rhodes
The University of Texas at Austin, Austin, TX, United States

Rizwan-uddin
Department of Nuclear, Plasma, and Radiological Engineering, Urbana; University of Illinois at Urbana-Champaign, Champaign, IL, United States

Piyush Sabharwal
Idaho National Laboratory, Idaho Falls, ID, United States

Sudhansu Sahoo
College of Engineering, Bhubneshwar, Orissa, India

Xiaohui She
Birmingham Centre for Energy Storage, School of Chemical Engineering, University of Birmingham, Birmingham, United Kingdom

F. Todd Davidson
The University of Texas at Austin, Austin, TX, United States

Lige Tong
School of Energy and Environmental Engineering, University of Science and Technology Beijing, Beijing, China

Li Wang
School of Energy and Environmental Engineering, University of Science and Technology Beijing, Beijing, China

Michael E. Webber
The University of Texas at Austin, Austin, TX, United States

Graham Wilson
Department of Mechanical and Nuclear Engineering, Kansas State University, Manhattan, KS, United States

Qinghua Yu
Birmingham Centre for Energy Storage, School of Chemical Engineering, University of Birmingham, Birmingham, United Kingdom

Tongtong Zhang
Birmingham Centre for Energy Storage, School of Chemical Engineering, University of Birmingham, Birmingham, United Kingdom

Xiaosong Zhang
School of Energy and Environment, Southeast University, Nanjing, China

Xinjing Zhang
Institute of Engineering Thermophysics, Chinese Academy of Sciences, Beijing, China

PREFACE

Nuclear energy currently provides about 11% of the world's electricity produced by about 450 nuclear power plants. It has almost solely been harnessed to produce reliable baseload electricity, when it comes to its civilian use. In recent decade, the concerns on climate change and energy security have resulted in the addition of intermittent renewable power leading to instabilities in the grid and price fluctuations. With the ongoing policy changes around the world to increase the penetration of renewable energy and industrial emphasis on using cheaper natural gas production for electricity production, it is important for nuclear power providers and operators to envision and implement new approaches to utilize the energy from nuclear fission reactors. This book describes some of the relevant technologies for hybridization of nuclear energy with renewables and storing energy from nuclear power plants. The scope covers the integration methods for the implementation with industrial facilities to efficiently use thermal and electric power of the nuclear reactors.

The content of the book covers the main technologies for hybridization and storage of nuclear energy. Chapter 1 describes the economics of nuclear power including advanced plant designs and the influence of nuclear fuel cycle. Chapter 2 provides an overview of hybridization and integration concepts that can provide paradigm shift for nuclear energy uses. Chapter 3 shows the impact of load following, with wind and nuclear energy systems in conjunction meeting the grid requirements on reactor safety.

Chapter 4 presents hydrogen production as an alternative use of nuclear energy, and Chapter 5 evaluates the suitability of different energy storage technologies for their integration potential with nuclear power. Chapters 6, 7, and 8 present chemical, thermal, and compressed air energy storage, respectively.

The objective in developing this book was to introduce the integration of nuclear energy and renewables, energy storage methods, and implementation of the technologies to industrial systems. This book is expected to be an invaluable source of reference for all those working directly in this important and dynamic field of hybrid energy systems; for beginners; and for scientists, engineers, and educators involved in the quest for clean and sustainable energy sources.

ACKNOWLEDGMENT

Hitesh Bindra thankfully acknowledges the Big 12 Faculty Fellowship Award that helped in the development of this collaborative book. He is thankful for several thought-provoking discussions with nonelectric energy storage pioneers Reuel Shinnar (deceased 2011) and Charles Forsberg (MIT). He would like to thank his past students and coworkers Jacob Edwards, Daniel Gould, Pablo Bueno, and Ryan Anderson for their insightful work on the energy storage topics. He gratefully appreciates the support of his family—Mamta, Shree, and Ohm—and his parents.

Shripad T. Revankar is thankful to the following:

His former students and collaborators in hybrid nuclear energy system, particularly Seungmin Oh, Cheikhou Kane, and Nicholas R. Brown for their work on nuclear hydrogen system and coupled hybrid systems.

His family members; wife Jayashree; and children, Vedang, Sachit, and Pavan, for their continued support.

CHAPTER ONE

Economics of Advanced Reactors and Fuel Cycles

Kathryn D. Huff*
*University of Illinois at Urbana-Champaign, Urbana, IL, United States

1.1 INTRODUCTION TO NUCLEAR POWER ECONOMICS

This chapter first introduces fundamental concepts in nuclear power economics, including a brief overview of the nuclear fuel cycle, electricity pricing, and calculation of the levelized cost of electricity. The impacts of capital costs are then covered, with special attention paid to financing, technical maturity, regulatory costs, and uncertainty inherent in advanced reactor builds. Trends in costing and pricing covered in the previous chapter impact these economics and will be touched on here as well.

Classes of advanced reactors and their particular distinguishing features will be discussed, with particular focus on variations in fuel management and economics. Finally, the economics of various advanced fuel cycles will be introduced and compared. Finally, comparative economic features of various advanced reactors and their respective fuel cycles will be summarized.

> *An academic reactor or reactor plant almost always has the following basic characteristics: (1) It is simple. (2) It is small. (3) It is cheap. (4) It is light. (5) It can be built very quickly. (6) It is very flexible in purpose ('omnibus reactor'). (7) Very little development is required. It will use mostly 'off-the-shelf' components. (8) The reactor is in the study phases. It is not being built now.*
>
> *On the other hand, a practical reactor plant can be distinguished by the following characteristics: (1) It is being built now. (2) It is behind schedule. (3) It is requiring an immense amount of development on apparently trivial items. Corrosion, in particular, is a problem. (4) It is very expensive. (5) It takes a long time to build because of the engineering development problems. (6) It is large. (7) It is heavy. (8) It is complicated.*
>
> ***Admiral Hyman G. Rickover***

Advanced and conventional nuclear reactor costs are both dominated by capital investment and operation and maintenance costs. Fuel costs, on both

Storage and Hybridization of Nuclear Energy
https://doi.org/10.1016/B978-0-12-813975-2.00001-6

the front and back end of the fuel cycle, are a small part of overall costs for conventional fuel cycles, but can contribute more significantly in certain advanced fuel cycles.

Meanwhile the electricity market generally dictates revenue. For advanced reactors that boast very high operating temperatures or load following behavior, unconventional revenue strategies may improve the outlook for those technologies in current electricity markets.

In addition to costs and revenues, advanced reactor economics are complicated by their technology readiness. Abdulla et al. [1] have argued that advanced fission research and development funding in the United States has been ineffectual. In particular, nuclear innovation is hobbled because "annual funding varies fourfold, priorities are ephemeral, incumbent technologies and fuels are prized over innovation, and infrastructure spending consumes half the budget."

Notably, however, increased experience with nuclear reactor construction and deployment has failed to lower costs [1a, 2]. Pathologically, nuclear power construction costs actually increase in costs despite increased experience with technology maturity and *n*th of a kind impact.

1.1.1 The Nuclear Fuel Cycle

The nuclear fuel cycle is the path of nuclear fuel material from its start as raw ore in the Earth's crust, through its processing and transmutation in nuclear facilities, and to its ultimate disposal in a repository.

As shown in Fig. 1.1, the front end of the fuel cycle begins with mining of uranium (or thorium) deposits, followed by milling, conversion, enrichment, and fuel fabrication. Together, these processes can be completed in a matter of months or years. Each process contributes to the cost of fuel. The reactor itself contributes to the cost of electricity via fuel efficiency and power generation capacity as well as operations and maintenance costs. Advanced reactor types may represent dramatic differences in these efficiencies and capacities. Finally the back end of the fuel cycle also impacts costs since the spent fuel must be managed. In once-through fuel cycles, this management consists of storage and disposal while in advanced fuel cycles it may also include reprocessing. We will explore each of the components of nuclear power cost in later sections.

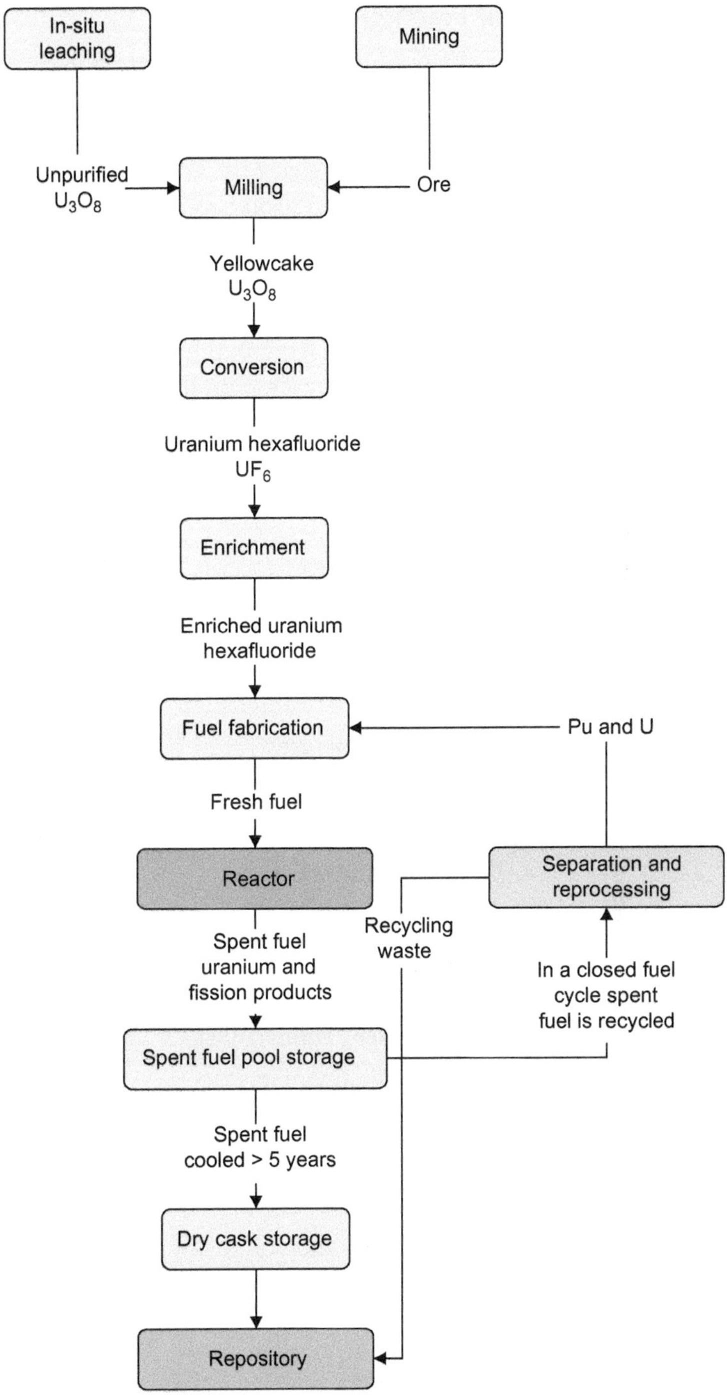

Fig. 1.1 A flowchart describing the nuclear fuel cycle. *Arrow* labels indicate the material form transferred between processes.

1.1.2 Levelized Cost of Electricity

The levelized cost of electricity (LCOE) is a way to measure holistically the costs, including the timeline of those expenditures, that go into the production of a kilowatt-hour. It is levelized over the lifetime of the plant. This measure is a useful, if imperfect, metric for electricity cost [3, 4] which incorporates the time value of money as well as variable profit and costs during the lifetime of a plant.

An easier calculation, very similar to the LCOE, is the approximate cost of electricity, which neglects this year-by-year assessment. The LCOE equation, as defined by Tsoulfanidis [3], has a large number of variables.

Values that must be calculated during each period, n are as follows.

$$E_n = \text{Energy sold} \tag{1.1}$$

$$I_n = \text{Investment outstanding at beginning of period} \tag{1.2}$$

$$Z_n = \text{Investment} \tag{1.3}$$

$$V_n = \text{Income from other than energy sale} \tag{1.4}$$

$$F_n = \text{Fuel expense} \tag{1.5}$$

$$D_n = \text{Depreciation} \tag{1.6}$$

$$O_n = \text{Operation and maintenance} \tag{1.7}$$

$$T_n = \text{Federal taxes} \tag{1.8}$$

$$S_n = \text{State taxes} \tag{1.9}$$

$$P_n = \text{Property tax} \tag{1.10}$$

$$G_n = \text{State gross sales tax} \tag{1.11}$$

In addition, there are variables that one may consider constant in time. These values do not vary during each period, n.

$$f = \text{Used fuel fee (1mill/kWe)} \tag{1.12}$$

$$N = \text{Life of investment in years} \tag{1.13}$$

$$\tau = \text{Federal tax rate} \tag{1.14}$$

$$s = \text{State tax rate} \tag{1.15}$$

$$g = \text{Gross sales tax rate} \tag{1.16}$$

$$p = \text{Property tax rate} \tag{1.17}$$

$$e = \text{LCOE} \tag{1.18}$$

$$f_b = \text{Fraction of investment in bond form} \tag{1.19}$$

$$f_s = 1 - f_b = \text{Fraction in form of stocks} \tag{1.20}$$

$$r_s = \text{Stock rate of return} \tag{1.21}$$

$$r_b = \text{Bond state of return} \tag{1.22}$$

Taxes factor into this as well.

$$T_n = \tau(10^{-3}eE_n + V_n - F_n - D_n - O_n - f_b r_b I_n - G_n - P_n - S_n - f \times 10^{-3}E_n) \tag{1.23}$$

$$S_n = \tau(10^{-3}eE_n + V_n - F_n - D_n - O_n - f_b r_b I_n - G_n - P_n - f \times 10^{-3}E_n) \tag{1.24}$$

$$G_n = g(10^{-3}eE_n + V_n) \tag{1.25}$$

$$P_n = pI_n \tag{1.26}$$

The sum of all these taxes is thus:

$$TT_n = \text{Total taxes} \tag{1.27}$$

$$= T_n + S_n + G_n + P_n \tag{1.28}$$

However, we can see that S_n and T_n differ only very slightly.

$$\begin{aligned} T_n + S_n = (\tau + s)(10^{-3}eE_n + V_n - F_n - D_n - O_n - f_b r_b I_n - G_n \\ - P_n - f \times 10^{-3}E_n) - \tau S_n \end{aligned} \tag{1.29}$$

To make the equation for total taxes simpler, we can define the combined state and federal taxes as just that, the combined state and federal taxes:

$$\tau_s = \tau + s$$

$$\begin{aligned} T_n + S_n = \tau_s(10^{-3}eE_n + V_n - F_n - D_n - O_n - f_b r_b I_n - G_n \\ - P_n - f \times 10^{-3}E_n) - \tau S_n \end{aligned} \tag{1.30}$$

Now we can get an equation for total tax:

$$TT_n = T_n + S_n + G_n + P_n \tag{1.31}$$

$$\begin{aligned} = \tau_s(10^{-3}eE_n + V_n - F_n - D_n - O_n - f_b r_b I_n - f \times 10^{-3}E_n) \\ - \tau S_n + (1 - \tau_s)g(10^{-3}eE_n + V_n) + (1 - \tau_s)pI_n \end{aligned} \tag{1.32}$$

We can simplify one step further by combining like terms:

$$\begin{aligned} TT_n = \tau_s\left(10^{-3}(e - f)E_n + V_n - F_n - D_n - O_n - f_b r_b I_n\right) \\ - \tau S_n + (1 - \tau_s)\left(g10^{-3}eE_n + gV_n + pI_n\right) \end{aligned} \tag{1.33}$$

The rest of the levelized cost is very similarly verbose. We define terms like the net income, R_n as derivatives of the previously defined terms.

$$R_n = \text{Net income} \tag{1.34}$$

$$\begin{aligned} &= (\text{Gross income}) \\ &\quad - (\text{Fuel disposal cost}) \\ &\quad - (\text{Operating and other costs}) \\ &\quad - (\text{Return on investment}) \\ &\quad - (\text{Debt payment}) \\ &\quad - (\text{Taxes}) \end{aligned} \tag{1.35}$$

This helps define the "investment outstanding":

$$I_{n+1} = I_n + V_n - R_n \tag{1.36}$$

Then, we seek to apply the effective cost of money to these outlays of money. The effective cost of money for these utilities factors in their stock and bond ratios, the returns on each, and the taxes under which they are burdened.

$$x' = r_s f_s + (1-\tau) r_b f_b \tag{1.37}$$

Without us going through the full derivation, note that Tsoulfanidis [3] comes up with:

$$e = \frac{\sum_{n=0}^{n} (1+x')^{-n} \left[\frac{Z_n - V_{n,nt}}{1-\tau} - V_{n,t} - \frac{\tau}{1-\tau} D_n + O_n + F_n + 10^{-3} f E_n \right]}{10^{-3} \sum_{n=1}^{N} (1+x')^{-n} E_n} \tag{1.38}$$

An easier calculation is the approximate cost of electricity, which neglects this year-by-year assessment, is:

$$e = 10^3 \frac{xI + O + F}{E}$$

where x is the annual fixed change rate (year^{-1}), I is the initial cost of the plant, O is operation and maintenance, F is annual fuel costs, and E is net electricity generated.

1.2 CAPITAL COSTS

Nuclear power capital costs, primarily necessary for reactor and fuel cycle facility construction, vary significantly internationally, but currently average approximately $10 billion in overnight construction costs [1]. This enormous outlay of funds requires significant financing, particularly since revenue. Accordingly, the actual cost of construction is very sensitive to financing rates, construction delays, and uncertainties.

Beyond construction costs, capital outlays required before reactor startup additionally include regulatory costs. These costs vary internationally and defy quantification, as regulation influences all components of reactor construction, operation, and maintenance.

1.2.1 Financing

When the cost of financing is included, many assessments [4] of nuclear power economics estimate that capital costs comprise over half of the resulting levelized unit electricity cost (also known as the levelized cost of electricity). In [5] levelized cost of electricity responds dramatically to changes in the discount rate (analogous to the cost of money). In that analysis, increasing the discount rate from 3% to 10% increases the levelized cost of electricity from an AP1000 from $32.56 to $61.20 per MWh. That is, increasing the discount rate by 7% could nearly double the cost of electricity generated by the AP1000.

1.2.2 Uncertainty

Nuclear reactors can take between 4 and 20 years to build. Over long time scale, appreciable uncertainty in financing rates, material pricing, and labor pricing increase the risk, and therefore the cost of investment. These uncertainties can require nuclear utilities to establish contingency strategies and maintain higher liquidity, both of which increase costs.

1.2.3 Delay

Delays in the construction process increase the length of time capital must be borrowed. By virtue of the time value of money, interest compounds as delays increase the time before revenue generation begins.

As a brief aside, the time value of money at the core of this issue should be stated mathematically. The *time value of money* means that a person places a higher value on a dollar in one point in time than on the same dollar at a different (usually later) point in time (see [3, 4]).

In mathematically describing the time value of money, we will use the following notation.

$$P = \text{Present value } (\$) \quad (1.39)$$
$$F = \text{Future value } (\$) \quad (1.40)$$
$$N = \text{Number of periods of investment (period)} \quad (1.41)$$
$$i = \text{Interest rate per period} \quad (1.42)$$

In the simplest case, the future value of money grows linearly with the number of periods invested.

$$F = P(1 + iN)$$

In a compounding interest scenario, the future value of money is *compounded* at each period. That is, after each period (e.g., each year) the total amount owed is note and then interest is applied to that amount. After the first period, the borrower is racking up additional interest on the interest that they owe. The equation is like this:

$$F = P(1 + i)^N$$

For the compounding interest form of $F(P, i, N)$ in the equation earlier, the future value of money (F) has an exponential relationship with time (N), as seen in Fig. 1.2.

At current financing rates, if a 10-year, $10-billion construction project is delayed by 1 year, the total cost can increase by hundreds of millions of dollars.

1.2.4 Technological Maturity

In most industries, initial attempts at commercializing new technologies are more expensive than later attempts. First-of-a-kind technology deployments typically impart lessons in efficiency, resource management, and process which can be carried through to later deployments (Nth of a kind). However, in nuclear power, this effect has not been seen. Instead, paradoxically, later deployments seem to grow in expense. Various explanations for this indicate that the conclusion to be drawn in the context of this discussion is simply that a typical economic model—in which first of a kind deployments cost less than later deployments—is unlikely to correctly model the economics of nuclear power deployments.

That said, the cost of research and development is significant for very new designs and can be difficult to project.

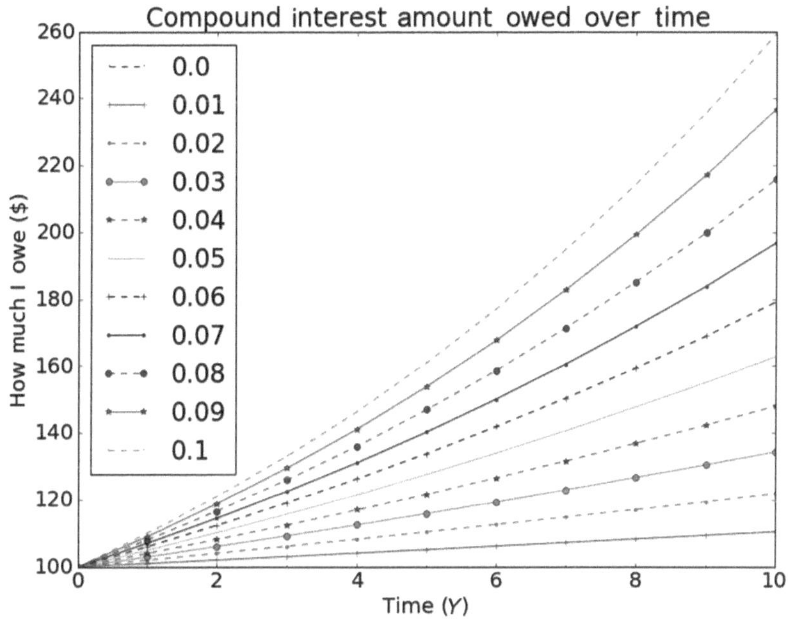

Fig. 1.2 Compounding interest, typical of most investments, gives an exponential relationship between the future value of money and time.

1.3 FRONT-END FUEL CYCLE COSTS

This section will explore the technologies and parameters involved in all of the contributors to fuel costs:

- Cost of raw uranium ($ per kgU)
- Cost of conversion ($ per kgUF6) (kgUF6/kgU)
- Cost of enrichment ($ per SWU) (SWU/kgU)
- Cost of transportation and fabrication ($ per kgU)

The amount of fuel needed is calculated in units of electricity (kWh) per kilogram (kg). It is the product of the following terms:

- Burnup (MWd(th)/t)
- Unit conversion (1000 kW(th)/MW(th))
- Unit conversion (10^{-3} t/kg)
- Unit conversion (24 h/day)
- Thermal-electric efficiency (kWh(e)/kWh(th))

1.3.1 Mining and Milling

Mining can take the form of open pit mining, underground mining, or in situ leaching. While open pit and underground mining each result in raw uranium ore with isotopic enrichment reflective of the natural abundance of various isotopes of uranium.[1]

The cost of mining varies slightly by method.

To purify mined uranium, a chemical process called milling concentrates and purifies the ore into U_3O_8, known as yellowcake. Uranium is sold in this form. The spot price of yellowcake has remained approximately $40/lb except for a price bubble in 2007, as shown in Fig. 1.3. That dramatic change in price is partly due to reduced surplus of stockpiled uranium and subsequent transition to newly mined uranium. That is, uranium production was at one point far ahead of demand, but recently, mining follows demand a bit more closely.

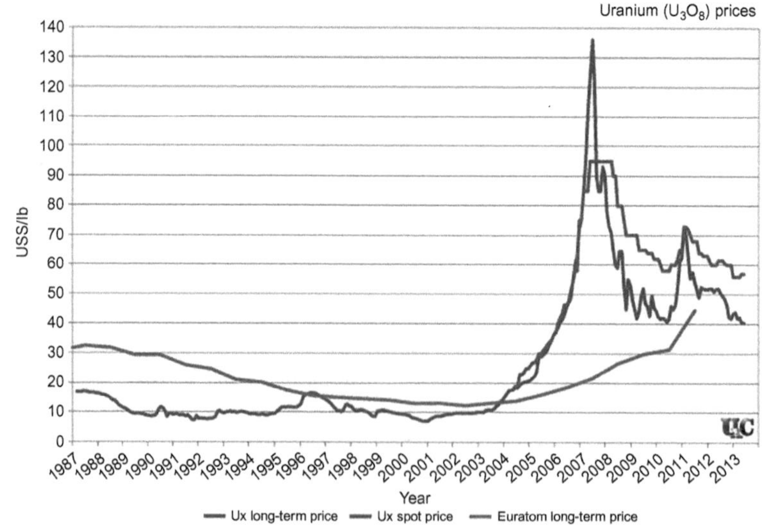

Fig. 1.3 The spot price and long-term price of uranium have varied over time.

[1] Natural uranium is predominantly ^{238}U with 0.0711% ^{235}U and trace quantities of ^{234}U.

Notably, one alternative option to mining the Earth's crust is mining the oceans for uranium through seawater extraction. While this method is prohibitively expensive at the moment, it may be competitive in the future and represents what is, for all realistic purposes, a renewable resource.

For some advanced reactors, the fertile isotope ^{232}Th serves to breed the fissile isotope ^{233}U. Importantly, even for these thorium-fueled designs, initial fissile uranium is required.

As far as availability is concern, a great deal of both uranium and thorium is available in the Earth's crust, depending on the price the market is willing to bear. Certain nations, such as Canada, Kazakhstan, and Australia, have significant natural uranium resources. In addition, our known resources continue to expand because prospecting for mineral resources and the ability to extract those resources respond to market price.

1.3.2 Conversion

In order to prepare uranium for enrichment, yellowcake (U_3O_8) must be *converted* to a gaseous phase uranium hexafluoride (UF_6). This conversion process is performed at high temperatures, exceeding 900°F. This step is necessary because most enrichment processes require that the material reaches a gaseous phase at relatively low temperatures. UF_6 is especially appropriate for enrichment because it is solid at room temperature, but gaseous at slightly elevated temperatures (at atmospheric pressure).

The conversion process includes a reduction step, a hydrofluorination step, and a fluorination step. In reduction, gaseous hydrogen, H_2, *reduces* U_3O_8, with a chemical balance that can be summarized as the following:

$$U_3O_8 + 2H_2 \rightarrow 3UO_2 + 2H_2O$$

This is followed by hydrofluorination, which is an interaction with anhydrous hydrogen fluoride (at 900–1000°F) that produces uranium tetrafluoride. The balanced form of this chemical reaction is thus.

$$4HF + UO_2 \rightarrow UF_4 + 2H_2O$$

Once the "hydrofluorination" step is performed, the uranium tetrafluoride (UF_4) is fluorinated into uranium hexafluoride (UF_6)

$$UF_4 + F_2 \rightarrow UF_6$$

Two main methods, dry hydrofluor and wet solvent extraction are used. The difference between these methods is when the purification step occurs (first or last).

Two major types of conversion are implemented internationally. These two processes have the same core features (reduction, hydrofluorination, and fluorination). But, they differ slightly. While the dry process ends with a purification step (distillation), the wet process begins with the purification step (solvent extraction and calcination).

Some losses in the process can increase the cost.

1.3.3 Enrichment

Enrichment is the stage at which natural uranium is processed to achieve higher concentrations of fissile ^{235}U isotopes (Fig. 1.4).

Uranium found in nature, U_{nat}, contains, by weight:

- 0.711% ^{235}U
- 99.284% ^{238}U
- Trace amount (0.0055%) of ^{234}U

The halflife of ^{238}U is longer than that of 235U, so the ratio of these two isotopes in natural uranium changes over the course of millions of years.

The physical effort required to enrich is related to the cost. Indeed, prices for enriched uranium production are typically quoted in dollars per *separative work unit (SWU)*. To understand the cost of enrichment, one must first understand SWUs. In order to define SWUs, the following terms are needed:

$$F = \text{Rate of feed material (per time)} \tag{1.43}$$
$$P = \text{Rate of enriched product material (per time)} \tag{1.44}$$
$$W = \text{Rate of waste material (tails)} \tag{1.45}$$
$$x_f = \text{Weight percent of the target isotope in feed} \tag{1.46}$$
$$x_p = \text{Weight percent of the target isotope in product} \tag{1.47}$$
$$x_w = \text{Weight percent of the target isotope remains in waste} \tag{1.48}$$

The core equations in the math of enrichment are these two:

$$F = P + W \tag{1.49}$$
$$x_f F = x_p P + x_w W \tag{1.50}$$

We can make these more useful by defining the following derived equations.

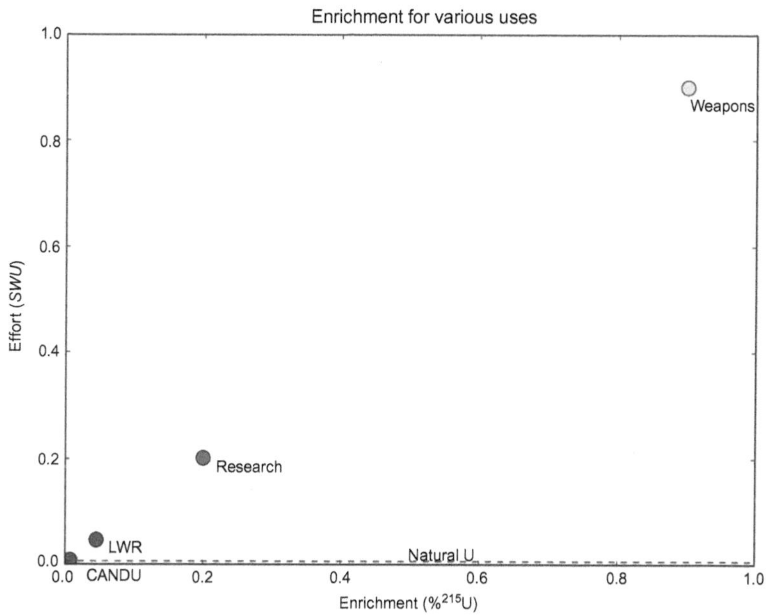

Fig. 1.4 ^{235}U enrichment typical for various applications.

$$FF = \text{The feed factor} \quad (1.51)$$

$$= \frac{F}{P} \quad (1.52)$$

$$= \frac{x_p - x_w}{x_f - x_w} \quad (1.53)$$

$$WF = \text{The waste factor} \quad (1.54)$$

$$= \frac{W}{P} \quad (1.55)$$

$$= \frac{x_p - x_f}{x_f - x_w} \quad (1.56)$$

In addition, by defining the separation potentials, $V(x_i)$, we can arrive at SWUs (Fig. 1.5),

$$V(x_i) = (2x_i - 1)\ \ln\left(\frac{x_i}{1 - x_i}\right) \quad (1.57)$$

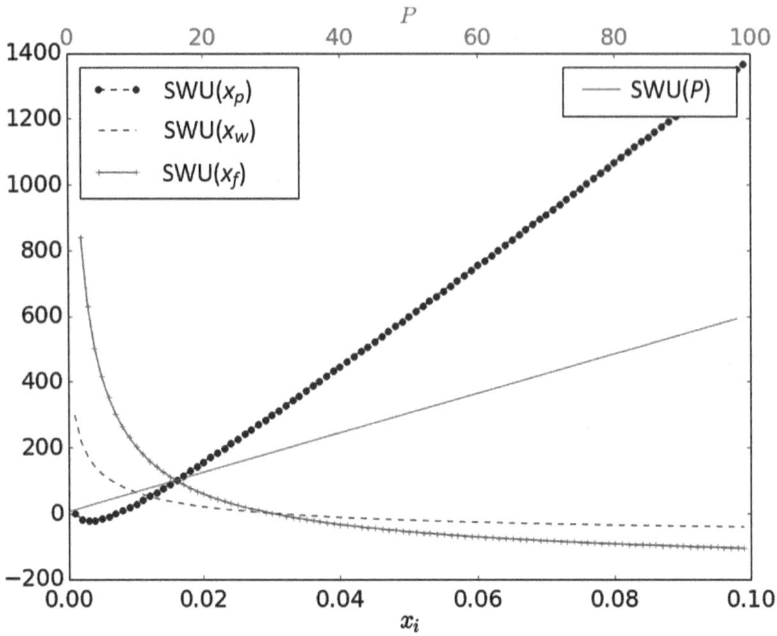

Fig. 1.5 While SWUs are a linear function of the rate of enriched product material, SWUs are a nonlinear function of the product enrichment, x_p, the waste enrichment, x_w, and the feed enrichment x_f.

$$\Rightarrow V(x_f) = (2x_f - 1)\ln\left(\frac{x_f}{1 - x_f}\right) \tag{1.58}$$

$$SWU = \tau(PV(x_p) + WV(x_w) - FV(x_f)) \tag{1.59}$$

1.3.4 Fuel Fabrication

Fabrication of nuclear fuel is a highly regulated, precise operation. International leaders in fuel fabrication include Areva and Westinghouse. From these fuel manufacturers, utilities can either lease or buy fuel. Ultimately, the cost of fabricating nuclear fuel varies by type. Advanced reactors have a variety of fuel types, which will influence the pricing.

Transportation of fresh nuclear fuel to the reactor can factor into the cost of fresh nuclear fuel as well.

1.4 BACK-END FUEL CYCLE COSTS

The cost of the back end of the fuel cycle can be challenging to quantify. The back end includes all elements of spent fuel management, storage, disposal, and optional reprocessing, as well as final decontamination and decommissioning of the reactor facility itself.

1.4.1 Reprocessing

The most common fuel cycle internationally is once-through, in which used nuclear fuel is destined for storage and disposal. However, significant valuable fissile material is still present in used nuclear fuel. Reprocessing is a process, which recovers uranium and plutonium isotopes from spent nuclear fuel (SNF). In this process, high-level waste (HLW) streams may be separated in preparation for special treatment before disposal.

France and Japan have both incorporated reprocessing of spent fuel into their nuclear fuel cycle.

By recycling U and Pu, substantial natural uranium savings can be achieved [3].

Let us introduce some new variables:

$$x_s = \text{Enrichment of uranium in used fuel} \tag{1.60}$$

$$fPu = \text{Mass of fissile plutonium} \tag{1.61}$$

$$s = \text{kg of fPu recovered per kg of used fuel} \tag{1.62}$$

$$u = \text{kg of used fuel per kg of fresh fuel} \tag{1.63}$$

We can calculate the savings in natural uranium due to plutonium recycling.

Consider 1 kg of fresh fuel. It contains x_p kg of ^{235}U. If we replace this kilograms with MOX (Pu and depleted U tails) we calculate an "equivalence":

$$x_p \text{ kg of } ^{235}\text{U} \leftrightarrow x_w \text{ kg of } ^{235}\text{U} + \frac{x_p - x_w}{p} \text{ kg of Pu} \tag{1.64}$$

So, if natural uranium is used as feed, FF kg of natural uranium are needed for each kg. If MOX is used instead, the savings is:

$$\frac{FF \cdot p}{x_p - x_w} = (\text{kgU saved per kg fPu}) \tag{1.65}$$

We can also describe savings in terms of s (Pu recovered) and amount of fuel discharged:

$$\frac{s \cdot FF \cdot p}{x_p - x_w} = (\text{kgU saved per kg of fuel discharged}) \tag{1.66}$$

Or, we can describe it per kilograms initially loaded into the reactor:

$$\frac{u \cdot s \cdot FF \cdot p}{x_p - x_w} = (\text{kgU saved per kg of fuel initially charged into the reactor}) \tag{1.67}$$

Finally, we can describe it per kilograms of natural uranium

$$\frac{u \cdot s \cdot p}{x_p - x_w} = (\text{kgU saved per kg of natural uranium feed}) \tag{1.68}$$

1.4.2 Storage

Storage of SNF begins in spent fuel pools where nuclear fuel cools. Following that the SNF is loaded into dry casks and stored on concrete pads near the originating reactor facility. This cost is currently absorbed by the facilities, which produced the fuel and will continue to require funding until a final disposal option is established.

1.4.3 Disposal

Permanent disposal of SNF is on hold in the United States, as the Yucca Mountain project continues to be mired in legislative limbo. The likelihood of licensing and operating a permanent spent fuel repository at Yucca Mountain is low. The only HLW disposal facility in the United States is the Waste Isolation Pilot Plant in New Mexico, which does not accept SNF from the US Power reactors. Beginning in 1982, the US Department of Energy collected a waste fee of 0.1 cents/kWh generated by nuclear power to fund management of the spent fuel. However, in 2014, the nuclear industry succeeded in petitioning to halt the collection of that fee, for as long as action toward a spent fuel repository continues to linger in indefinite limbo. This fund has been earning interest. While the Nuclear Energy Institute calculates that the fund was worth $42.8 billion in 2014, the Department of Energy estimates $30B. Thus, between 1982 and 2014, the cost of SNF disposal was a flat 1 mill/kWh. The cost of disposal may be less clear cut in the future, pending further legislation.

The situation internationally is slightly more hopeful. While most nuclear power nations have no plan for permanent disposal, Sweden and Finland have both moved forward. Each has successfully sited their future repository and are pushing forward with construction.

1.4.4 Decommissioning

In the United States, a fund for the decommissioning of the reactor facility is established at the beginning of operation and is available throughout the reactor lifetime. Accordingly in this chapter, decommissioning is captured in the initial capital costs.

1.5 ADVANCED REACTORS

The Generation IV International Forum identified numerous reactor designs of potential serious consideration internationally. Recently, a plethora of advanced reactor designs is in various stages of research and commercialization.

This section is informed by the following references [2, 5–9].

1.5.1 Classes of Reactors

Nuclear reactors vary enormously and can be classified along many axes. The energy of the neutron spectrum can be "fast" or "thermal." This can result in either breeding or burning the fissile material over time.

Along another axis, the fuel can be solid or liquid. Among solid fuels, fuel geometry may be pins, plates, tristructural isotropic (TRISO) microkernels, and quadstructural isotropic (QUADRISO). These microkernels can be embedded in pebbles, prisms, or pins.

In addition, the removal of heat can be conducted by many types of coolant including water, sodium, lead, helium gas, or molten salts.

Some reactors combined the coolant with the fuel. Liquid fuels typically take the form of salts. Typically uranium or thorium is dissolved in fluoride, chloride, or uranyl salts.

The following sections highlight a few reactor classifications and the key economic implications of design decisions inherent to those designs.

1.5.2 High-Temperature Reactors

Reactors, which operate at very high temperatures, such as the high-temperature gas reactor or various fluoride cooled high-temperature reactors could improve economic feasibility through diversification. That is, very high-temperature heat is useful for various types of chemical and industrial processing (desalination, hydrogen production, steel manufacturing, etc.) The higher temperatures produced by these reactors could offer a cheaper, emissions-free alternative to current heat-generating stations at industrial facilities, which consume significant natural resources for heat generation.

1.5.3 Liquid-Fueled Molten Salt Reactors

Liquid-fueled reactors, particularly the molten salt kind, promise exceptional fuel utilization and sometimes very high operating temperatures. However, their most transformative inherent advantage is the potential for load following. Because problematic fission products, such as xenon 135, can be removed from liquid fuels during reactor operation, these reactors have the potential to rapidly respond to electrical demand on the grid.

1.5.4 Breeder Reactors

While some reactors primarily burn fissile material, others can generate fissile material through neutron capture in fertile isotopes. These reactors are called breeder reactors and boast impressive fuel utilization. Breeder reactors are often fast spectrum uranium plutonium reactors like the sodium cooled fast reactor or the liquid metal-fueled breeder reactor. They can also be high-temperature, liquid-fueled molten salt reactors (like the molten salt breeder reactor (MSBR)). The benefits of these reactors are primarily in significantly improved fuel utilization. While this improvement has important environmental consequences, the economic ones are not so clear, since fuel is a very small part of nuclear fuel costs in general.

1.6 ADVANCED FUEL CYCLES

This section will include discussion informed by the following references [10–14]. Advanced fuel cycles promise to improve spent fuel management in two main ways.

First, reprocessing fuel cycles typically promise smaller volumes of spent fuel, often with shorter-lived waste forms. Notably, while shorter-lived waste is appealing from a long-term management perspective, short half lives are analogous with higher activity. Accordingly, these waste forms can be hotter than waste forms from conventional fuel cycles. High heats can constrain design choices in storage and disposal.

Alternatively, some advanced reactor types promise a contained waste form instead. Tristructural isotropic (TRISO) fuel particles, embedded in either graphite pebbles or prisms, promise to serve as their own waste form.

Beyond this, advanced fuel cycles have costs modeled similar to existing fuel cycles.

1.7 SUMMARY

Myriad parameters impact nuclear reactor and fuel cycle costs. While policies such as carbon taxes, renewable portfolio standards, research and development funding, and other market interventions may impact electricity pricing and influence nuclear power revenues at a fundamental level, costs significantly impact the equation. In current reactors and their associated fuel cycles, the cost of money, particularly with regard to construction capital and operation and maintenance increasingly dominate the economics of nuclear power. Advanced reactors may promise new approaches to the market, such as providing high temperatures for industrial applications, incorporating hybridization, or enabling load following.

REFERENCES

[1] A. Abdulla, M.J. Ford, M.G. Morgan, D.G. Victor, A Retrospective Analysis of Funding and Focus in US Advanced Fission Innovation, Environ. Res. Lett. 12 (8) (2017) 084016. https://doi.org/10.1088/1748-9326/aa7f10. Available at http://stacks.iop.org/1748-9326/12/i=8/a=084016.

[1a] J.R. Lovering, A. Yip, T. Nordhaus, Historical construction costs of global nuclear power reactors, Energy Policy 91 (2016) 371–382.

[2] C.P. Pannier, R. Skoda, Comparison of small modular reactor and large nuclear reactor fuel cost, Energy Power Eng. 6 (5) (2014) 82.

[3] N. Tsoulfanidis, The Nuclear Fuel Cycle, American Nuclear Society, La Grange Park, IL, 2013.

[4] Lazard, Lazard's levelized cost of energy analysis ("LCOE") Version 10.0, (Financial Advisory and Asset Management No. 10.0), Perspective: Lazard's Levelized Cost of Energy Analysis, Lazard, New York, NY, 2016.

[5] L. Samalova, O. Chvala, G.I. Maldonado, Comparative economic analysis of the integral molten salt reactor and an advanced PWR using the G4-ECONS methodology, Ann. Nucl. Energy 99 (2017) 258–265.

[6] S.K. Kim, W.I. Ko, Y.H. Lee, Economic viability of metallic sodium-cooled fast reactor fuel in Korea, Sci. Technol. Nucl. Ins. 2013 (2013) e412349.

[7] G. Locatelli, C. Bingham, M. Mancini, Small modular reactors: a comprehensive overview of their economics and strategic aspects, Prog. Nucl. Energy 73 (2014) 75–85.

[8] Y.I. Chang, The integral fast reactor, Nucl. Technol. 88 (2) (1989) 129–138.

[9] N.R. Brown, A. Worrall, M. Todosow, Impact of thermal spectrum small modular reactors on performance of once-through nuclear fuel cycles with low-enriched uranium, Ann. Nucl. Energy 101 (2017) 166–173.

[10] D.E. Shropshire, K.A. Williams, W.B. Boore, J.D. Smith, B.W. Dixon, M. Dunzik-Gougar, R.D. Adams, D. Gombert, Advanced fuel cycle cost basis. Idaho National

Laboratory, 2007https://doi.org/10.2172/911948. Technical Report OSTI ID: 911948, INL/EXT-07-12107.

[11] R. Gao, S. Choi, W. Ko II, S. Kim, Economic potential of fuel recycling options: a lifecycle cost analysis of future nuclear system transition in China, Energy Policy 101 (2017) 526–536.

[12] J.B. Greenblatt, N.R. Brown, R. Slaybaugh, T. Wilks, E. Stewart, S.T. McCoy, The future of low-carbon electricity, Annu. Rev. Environ. Resour. 42 (1) (2017) 289–316.

[13] S. Fetter, M. Bunn, J.P. Holdren, B. van der Zwaan, The economics of reprocessing vs. direct disposal of spent nuclear fuel, (2003). http://drum.lib.umd.edu/handle/1903/4043 (accessed 01 March 2018).

[14] W. Stanek, J. Szargut, Z. Kolenda, L. Czarnowska, Exergo-ecological and economic evaluation of a nuclear power plant within the whole life cycle, Energy 117 (Part 2) (2016) 369–377.

CHAPTER TWO

Hybrid and Integrated Nuclear Power, Compressed Air Energy Storage, and Thermal Energy Storage System

Rizwan-uddin
Department of Nuclear, Plasma, and Radiological Engineering, Urbana, IL, United States
University of Illinois at Urbana-Champaign, Champaign, IL, United States

2.1 INTRODUCTION

2.1.1 Nuclear Power

While Gen-IV nuclear reactors—with their higher exit temperatures and thus higher quality of energy—are expected to convert a higher fraction of the fission (thermal) energy to electricity, both Gen-II and Gen-III reactors are largely restricted to exit temperatures that range between 290°C and 340°C. This restricts their conversion efficiencies to below 35%, and thus, a very large fraction (~65%) of the thermal energy produced in these plants is "wasted." Gen-II (and to a large extent, Gen-III) reactors were designed under economic conditions in which electricity was expensive, and thus, the sale (in the form of electricity) of just one-third of the energy produced in fission was sufficient to break even. However, recent growth of shale gas, growth of (and subsidies for) solar and wind power, increase in operation and maintenance (O&M) cost of nuclear power, and constantly improving device efficiencies (that lead to slower increase in power demand) are making it necessary that for nuclear power plants (NPPs) to compete in the deregulated market place, they must be operated more efficiently. While efforts are underway on all fronts to make nuclear power more competitive, for example, by reducing the O&M cost, major advances can be made by making better use of the energy from NPPs that is currently being released into the environment. An example of such an effort is the recent study to cogenerate electricity and liquid fuels, albeit using high-temperature gas reactors [1].

Storage and Hybridization of Nuclear Energy
https://doi.org/10.1016/B978-0-12-813975-2.00002-8

While utilization of waste thermal energy is one way to improve the efficiency of NPPs, another approach would be to make better use of the primary thermal energy that a baseload NPP produces. NPPs were designed as baseload electricity generators—that is, they were not expected to change their power level in response to grid demand. With the introduction of an increasing fraction of solar and wind power into the electricity grid—sources that are intermittent and, in many cases, subsidized—electricity prices in deregulated markets fluctuate significantly. A power source with subsidies can sell electricity to the grid at negative price and still be profitable. At times when electricity price is negative, baseload providers that cannot simply turn production off, such as NPPs, need to *pay* to off-load their electricity. Efforts are underway to explore the extent at which NPPs can operate in load-following mode. While reducing the power being produced will help when the electricity prices are negative, it is not the most efficient use of the capital that is invested in a NPP.

Hence, in addition to the load-following capability, advances in two additional areas are needed to address the challenges being faced by Gen-II and Gen-III reactors:

- Monetization of the waste heat from NPPs
- Coutilization or storage of energy (either thermal or electric) produced in NPPs during full- or partial-capacity operation

For example, under certain market conditions, utilization of the thermal energy produced by NPPs as process heat may be a better value proposition than converting that thermal energy to electricity. The more hybrid and integrated the production, storage, and reuse of energy is, the more efficient the overall system is likely to be. It is realized that significant gain in efficient use of energy sources will result from integration and hybridization that will lead to minimal waste and allow efficient production, transmission, storage, and consumption. Two major factors in the evaluation of almost all such propositions are technology readiness and cost-effectiveness.

After introducing the components of one such hybrid system, this chapter will describe different integration methodologies and assess potential gains.

2.1.2 Waste Energy From Nuclear Power

Our discussion here is largely restricted to pressurized water reactors (PWRs) and boiling water reactors (BWRs). Extension to other designs is in some cases trivial but may require additional discussion for yet different designs of NPPs.

As mentioned above, about 65% of thermal energy produced in an NPP is released into the environment. It is low-quality energy, but there are applications such as water desalination, district heating, fish farming, and agricultural uses in which this low-quality thermal energy can be effectively utilized. There is a growing body of literature reporting analyses and applications of such efforts. Focus of this chapter is only on one aspect of utilization of NPP's waste heat, namely, its potential use in a compressed air energy system.

2.1.3 Energy Storage

Storage—or, more precisely, grid-scale storage—is the holy grail in the energy field. Energy storage can take different forms—chemical, potential, kinetic, mechanical, etc. Some of these forms have specific georequirements such as a hill or a mountain for storing energy in potential form or a cave. These are usually based on fairly mature technologies. Other forms of large-scale energy storage still depend on new scientific and technical advances to become feasible.

2.1.4 Reuse of Stored Energy

Reuse of stored energy is in some cases fairly straightforward. For example, energy stored in pumped hydro system can be easily converted into a usable form (electricity) with reasonably fast control. Such systems can be readily integrated in most hybrid energy networks. On the other hand, integration of energy stored in thermal, mechanical, or chemical forms still requires some scientific and technological advances both in storage and in reuse stages. Examples of energy storage in the form of hydrogen and thermal energy are discussed elsewhere (Chapters 4 and 7). This chapter is aimed at the design of a hybrid system in which thermal and mechanical (compressed air) forms of stored energy are to be integrated. Individual forms of these energy storage modes are described first in the next section.

As noted elsewhere in this chapter, compressed air energy storage at grid-scale is limited by the availability of natural underground reservoirs. Though not the topic of this chapter, liquid air energy storage (LAES) also called cryogenic energy storage (CES) overcomes this particular drawback of CAES. It leads to much higher energy density and has no geographic constraints. In LAES, the energy is stored in a cryogenic fluid in liquid phase. It therefore greatly reduces the volume of the reservoir. LAES consists of an air liquefaction unit and an energy extraction unit. LAES are not further discussed in this chapter.

2.2 ENERGY STORAGE AND REUSE TECHNOLOGIES

As described above, there are numerous energy storage and reuse technologies. We here will focus on two of these that are part of the integrated, hybrid system proposed in this chapter—thermal and compressed air.

2.2.1 Thermal Energy Storage

Thermal energy storage promises to be the most cost-effective approach for energy storage. It is largely based on existing proved technologies. Advances in thermal energy storage are often in exploring suitability of new materials. One specific form of thermal energy storage (packed beds) is discussed in sufficient detail in Chapter 7.

Since different options available for thermal energy storage (e.g., solid vs liquid) are discussed in Chapter 7, we will skip them here.

2.2.2 Compressed Air Energy Storage System

Compressed air energy storage (CAES) systems were historically proposed, developed, and analyzed in the context of intermittent sources of energy, such as solar and wind. Goal was to increase the capacity factor and to improve economic feasibility of these energy sources for local- or grid-scale energy storage [2–6]. Excess electric energy available is stored in the form of compressed (high pressure) air in large spaces, such as tanks or salt caverns. Compressors are usually used to compress air. Energy reuse is usually by driving a turbine using stored compressed air. CAES clearly depends upon the availability of very large "containers" that can hold high-pressure air. Using existing technologies, next to pumped hydro, CAES systems promise to be the second most viable option for large- or grid-scale energy storage. Though energy storage in different modalities of compressed air has seen its advantages and disadvantages [3] and is not yet commercially viable, it has shown significant advances in recent years due to recent demands in energy storage. An indicator of its potential is venture capital start-up companies aiming to commercialize CAES. There is clearly room for a few major breakthroughs and resolution of technical challenges and significant room for optimization of design and coupling strategies with existing production, consumption, transportation, and grid connection. Major or minor breakthroughs and optimizations will likely make compressed air a strong contender in the crowded field of rapidly maturing energy storage and conversion technologies.

Some studies have suggested compressed air to be the leading technology for large-scale energy storage [7]. The idea has been around for a long time [8]. Germany built a 290 MW compressed air energy system (CAES) in 1978. Only unit operating in the United States is in Alabama (110 MW). Both of these store compressed air in excavated salt caverns. Storage pressure is around 70 bar, similar to the pressure in a boiling water reactor (BWR). In addition to the conventional candidates, depleted natural gas fields can also be potential candidates for storage. PNNL has studied the potential use of more commonly available porous and permeable rock formations. They showed that these other formations not only are suitable for CAES but also in some cases have advantages over salt mines [9]. Though PNNL study was carried out in the context of wind power (in the Pacific Northwest), results can be extended to nuclear power.

As in other energy storage systems, key parameters for technical viability of CAES are time and rate of loading and the total time and rate of discharge. A design specification may require the CAES to provide 100 MW of sustained power for, say, 50 h. For a tank-based CAES system, typical specs might be in the order of a few megawatts for, say, 10 h. Due to low energy density, large volumes are needed for compressed air storage. To relax the large-volume requirement, breakthroughs are needed to improve energy density of the storage system.

Compressed air energy storage systems are usually of two kinds: nonadiabatic and adiabatic. These are described below.

2.2.3 Compressed Air Energy System—Nonadiabatic

Currently operating CAES plants are of nonadiabatic kind in which the heat generated in the compression process is lost to the environment and needs to be supplied back (say, by burning natural gas) when energy needs to be recovered. This reheating stage normally uses natural gas, thus decreasing efficiency and releasing CO_2 into the atmosphere. This leads to a recovery in nonadiabatic systems of only around 50%. Luo et al. [10] reviewed the state of the art till 2014. A schematic diagram of a conventional CAES system is shown in Fig. 2.1 [10]. If the heat of compression can be stored at economically feasible cost, it can lead to significant improvement in CAES viability.

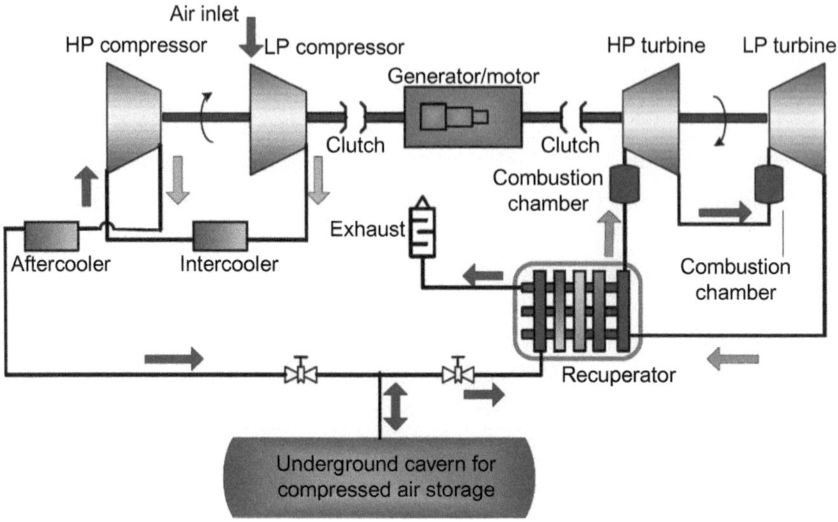

Fig. 2.1 Schematic diagram of a conventional CAES. *(From X. Luo, J. Wang, M. Dooner, J. Clarke, C. Krupke, Overview of current development in compressed air energy storage technology, Energy Procedia 62 (2014) 603–611.)*

2.2.4 Compressed Air Energy System—Adiabatic

In the *adiabatic* CAES, the heat generated during the compression phase is to be "recovered and saved," and it is this heat that is to be supplied back during the expansion phase in the combustors, thus potentially increasing the recovery to over 70%–75%. Thermal oil and molten salt are among the candidates for storing the thermal energy. Packed-bed thermal energy storage systems may also be viable. A prototype to demonstrate the technical and economic feasibility of an adiabatic system is being developed in Germany. General Electric is a major partner in this effort. There is significant room for R&D and optimization of the adiabatic approach. Due to low temperatures, quality of the energy extracted from the hot compressed air is not very high.

2.3 HYBRID AND INTEGRATED PRODUCTION, STORAGE, AND REUSE

Referring to Fig. 2.1, one can see that the thermal energy generated in the compression process is lost in the intercooler and aftercooler. Moreover, energy is needed to reheat the (cold) compressed air before it enters the HP turbine. In currently operating CAES, this is normally accomplished by burning natural gas (combustion chamber in Fig. 2.1).

A hybrid and integrated (CAES and thermal storage) system that can be used with nuclear power is described here. Such a hybrid and integrated system is explained in an expanded version of Fig. 2.1, shown in Fig. 2.2. The three components of this integrated system are an NPP, a CAES, and a thermal energy storage (TES) system. NPP is considered to be a standard PWR. Thermal energy storage system is similar to the one described elsewhere in this book (Chapter 7). Letters **B**–**G** are used to identify the flow of electric or thermal energy between different parts of the three components. **B** represents the flow of thermal energy from the CAES to the TES system as the thermal energy is released in the compression process. This energy, which in

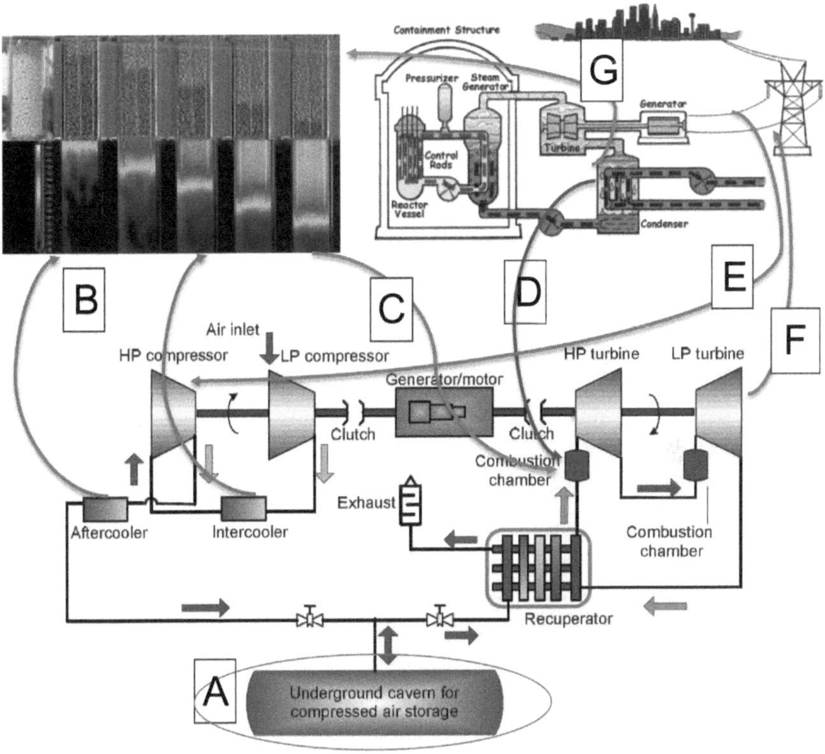

Fig. 2.2 Schematic diagram of a CAES [10], an NPP, and a packed-bed energy storage system (top left). Key components of this integrated system to be studied in this project are identified as **A–D** and **G. E** and **F**, respectively, represent the flow of electric energy from the NPP (or the grid) to the CAES to run the compressor and from the CAES to the grid. *From Fig. 2.2 B is published in the article: E. Jacob, H. Bindra, An experimental study on storing thermal energy in packed beds with saturated steam as heat transfer fluid,* Solar Ener. *157 (2017) 456–461.*

a nonadiabatic system is released to the atmosphere, in this hybrid and integrated system will be stored in the TES system. **G** represents the thermal energy from the NPP that is otherwise wasted, flowing to the TES system. It is likely that the TES system will need to be composed of two different configurations—one optimized to store the thermal energy from the NPP and the other optimized to store the thermal energy from the CAES. **A** is used to identify the CAES. **C** and **D**, respectively, represent the flow of thermal energy from the TES system and the NPP, to reheat the compressed air in the CAES before it is used to drive the turbines. **E** and **F**, respectively, represent the flow of electric energy from the NPP (or the grid) to the CAES to run the compressor and from the CAES to the grid. Key objectives in the design of any such hybrid system are to assess the links needed between the three subsystems, identified by letters **B**–**G**. Optimized designs of these links will constitute a major part of any study of such a hybrid system.

2.4 CHALLENGES AND GAINS

Among the challenges associated with this system, the foremost is the identification and assessment of geologic formations that exist near or around NPPs. These formations should be assessed for their suitability to store compressed air. Such studies require permeability and porosity to evaluate the storage capacity and other geologic conditions to assess suitability of the site to store compressed air.

In an adiabatic CAES system, thermal energy generated in the process of air compression is recovered and stored, for later use in the energy recovery stage. Because of low temperatures involved, the packed-bed TES system described in another chapter is an ideal candidate to store this energy. This TES system has been shown to be suitable for energy storage at even low temperatures [11]. Technical details of the thermal energy transfer from the compressed air to the packed-bed TES system have not been worked out.

Rather than simultaneously storing the thermal energy from the NPP in the TES system and simultaneously withdrawing thermal energy from the TES system to heat up the compressed air in the CAES before it enters the turbine (**G** and **C** in Fig. 2.2), it is likely to be more efficient—though challenging to design—to take the thermal energy directly from the NPP and use it to heat the compressed air (**D**). This will eliminate the thermal energy storage step. This may become even more important if the low quality of the thermal energy available in the compression stage of the CAES system makes its storage not feasible. Another challenge in the design of this hybrid and integrated system is how to design the turbine for optimum use of pressurized air (with some variations in inlet pressure and flow) and its associated heated system.

Technical challenges in designing the system to compress and store air are as follows:

- Discharge
- Heat storage
- System integration
- Optimization for economic feasibility
- Economic analysis
- Loading and discharge rates
- Safety

Building on existing strengths in three areas—namely, CAES, NPP, and thermal energy storage—an integrated design of the three subcomponents that will lead to a much better energy utilization than is possible in standalone form.

REFERENCES

[1] M.W. Patterson, Cogeneration of electricity and liquid fuels using a high temperature gas-cooled reactor as the heat source, Nucl. Eng. Des. 329 (2018) 204–212.

[2] R.H. Byrne, M.K. Donnelly, V.W. Loose, D.J. Trudnowski, Methodology to determine the technical performance and value proposition for grid-scale energy storage systems: a study for the DOE energy storage systems program, Sandia report, SAND2012-10639, December, 2012.

[3] R.H. Schulte, N. Critelli Jr., K. Holst, G. Huff, Lessons from Iowa: development of a 270 megawatt compressed air energy storage project in midwest independent system operator, a study for the DOE energy storage systems program, Sandia report, SAND2012-0388, January, 2012.

[4] A. Cavallo, Controllable and affordable utility-scale electricity from intermittent wind resources and compressed air energy storage (CAES), Energy 32 (2007) 120–127.

[5] P. Denholm, R. Sioshansi, The value of compressed air energy storage with wind in transmission-constrained electric power systems, Energy Policy 37 (8) (2009) 3149–3158.

[6] E. Drury, P. Denholm, R. Sioshansi, The value of compressed air energy storage in energy and reserve markets, Energy 36 (8) (2011) 4959–4973, https://doi.org/10.1016/j.energy.2011.05.041.

[7] Packing Some Power, The Economist, Q1, 2012 http://www.economist.com/node/21548495. http://energystorage.org/compressed-air-energy-storage-caes.

[8] P. Zaugg, Air-storage power generating plants, Brown Boveri Rev. 62 (1975) 338–347.

[9] B.P. McGrail, et al., Techno-economic performance evaluation of CAES in the Pacific Northwest, Report PNNL-22235, 2013.

[10] X. Luo, J. Wang, M. Dooner, J. Clarke, C. Krupke, Overview of current development in compressed air energy storage technology, Energy Procedia 62 (2014) 603–611.

[11] H. Bindra, P. Bueno, J.F. Morris, R. Shinnar, Thermal analysis and exergy evaluation of packed bed thermal storage systems, Appl. Therm. Eng. 52 (2013) 255–263.

CHAPTER THREE

Nuclear-Wind Powered Microgrid

Reduced-Order Modeling of LWR Response

Abhinav Gairola*, Hitesh Bindra*
*Department of Mechanical and Nuclear Engineering, Kansas State University, Manhattan, KS, United States

With four parameters I can fit an elephant, and with five I can make him wiggle his trunk.

John Vonn Neumann

3.1 INCREASING GRID PENETRATION OF RENEWABLE ENERGY SOURCES

The cost of electricity from solar photovoltaic (PV) systems has dropped 15- to 20-fold over the last 20 years, and the LCOE for a PV system currently is about $70–130 per MWh, including support structures, power conditioning, and land. However, these PV systems cannot be used to supply peak load for most of the locations due to time lag in peak electricity generation and peak demand. Moreover, an increased penetration of solar generation on the grid can have unintended consequences as it can increase the load-generation unbalance when weather conditions abruptly change. Similarly, onshore wind generation is another method of electricity production which can be considered as robust and economical but not reliable. If the grid is solely depended on solar PV cells or wind energy for a large fraction of the electricity supply, the back-up power required on the grid could equal or even exceed 50% of power consumed because the back-up power would have to be at least equal to the amount of power that is actually used from PV cells/wind turbines. A stable grid, therefore, requires excess capacity, which comes with additional back-up power but it remains idle for almost 50% of the time. The reliability of these energy systems in supplying peak load and base load can be substantially improved if there existed economical and durable methods of storing electricity, But the means of energy storage are currently not economical for grid scale deployment and therefore grid acts like a storage with the help of peak demand prices of electricity. Due to these reasons, the cost of electricity to consumers goes up when

https://doi.org/10.1016/B978-0-12-813975-2.00003-X

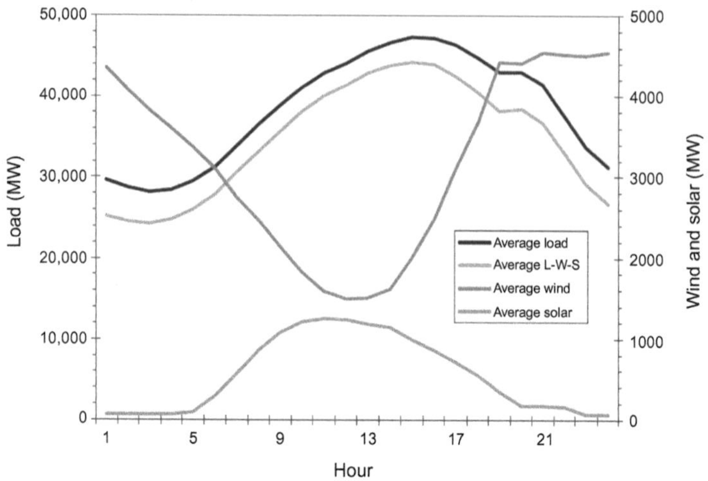

Fig. 3.1 A typical curve of wind and solar output, along with net (L-S-W) demand. *L*, load; *S*, solar; *W*, wind (NERC [1]).

PV units or wind energy without storage supply energy to the grid. Therefore, under all possible conditions the fractional share of renewables has remained or will remain up to 10–15% in the near future.

In an unregulated electrical grid, wholesale electricity prices are generally driven by the supply and demand, which depends on, time of day, weather conditions, seasonal factors, and consumption behaviors. Thus prices fall with low demand and increase during the high demand period. In the current market scenario, government policies lead to much higher revenues for producing electricity from wind and solar energy when compared to the more traditional sources of generation. The typical solar and wind energy production around the United States and across the world does not contribute significantly to reduce the demand load throughout a 24-hour period as shown in Fig. 3.1.

3.1.1 Impact on Nuclear Power

The government policies are designed to increase the solar and wind energy penetration in the electrical grid and conventional energy generation is expected to be flexible enough to fill in the deficits to the grid load demand. Due to this, some of the conventional power providers are even getting marred by negative electricity prices, where they have to pay the grid operators to offload the power they generate. Nuclear power plants (NPPs) are often subjected to negative pricing if they dispatch more energy on the grid than the net demand [2, 3].

In 2016, Bloomberg New Energy Finance published a report—"Reactors in the Red: Financial Health of the US Nuclear Fleet," which shows that 55% of America's nuclear plants are losing money and are at serious risk of being replaced by other sources of energy. Owing to which some of the conventional power producers may have to pay the consumers to offload the power from the grid. Lack of grid scale energy storage system suggests that the only options available for NPPs to overcome these economic challenges are to follow the grid load demand and save themselves from potential collapse.

3.1.2 Load Following NPPs

NPPs were designed and generally operate as base load providers. Stringent safety criteria derived from conservative considerations imposed by the nuclear regulators make it difficult to operate NPPs in load following mode. Due to these reasons, load following, adopted so far in very few nuclear plants, is limited to some specific ramp rates [3] or for only limited number of power changes in a day. From fuel response to stability of NPP systems, numerous aspects must be carefully studied before the load following envelope can be expanded. In United States or other parts of the world, the stringent safety criteria derived from conservative considerations—from the fuel performance point of view does not allow power to be ramped up and down in a short period of duration. For example, California's sole remaining NPP, the Diablo Canyon, does not ramp down during the middle of the day, despite frequent and recurring situations of negative prices, thus losing valuable dollars and making it uncompetitive when compared to renewables. From the clad-pellet interaction point of view, a ramping rate of 50 MW/min (5% per minute) load increase is derived for some European reactors [4].

Due to these reasons, it is important to examine reactor response under a scenario in which the electrical grid abruptly demands an increase in power for a finite period of time before the demand drops back to its original level. As energy production from renewable sources can be associated with sudden fluctuations, it is also critical to examine the response of NPPs to provide back-up fluctuating loads. For illustrating this, a hypothetical microgrid is modeled which is connected to, and is supplied electricity from a wind energy generation system and a light-water cooled NPP. Time-series data for a typical wind energy generation system are used as an input parameter for estimating the demand power for the NPP. As wind energy generation is dependent upon instantaneous wind speed for any geographical location, the discrete time-series information are then used to compute nuclear system response. This work models the nuclear-coupled thermal hydraulic response of the reactor when its power level is governed by the grid demand. These transients were simulated using a simple reduced-order model of a pressurized water reactor described in the next section.

The scope of this chapter is to provide a framework to assess possible thermal dynamics of the reactor under different load demand scenarios. In this study, there are two example test scenarios considered to show the implementation of this framework: (a) scenario where there is a step change in the grid power demand for a finite amount of time, and (b) considering a small microgrid where an NPP and stochastic wind energy generators exists together leading to sudden power demand fluctuations on nuclear plant (Figs. 3.2 and 3.3).

www.nei.org.

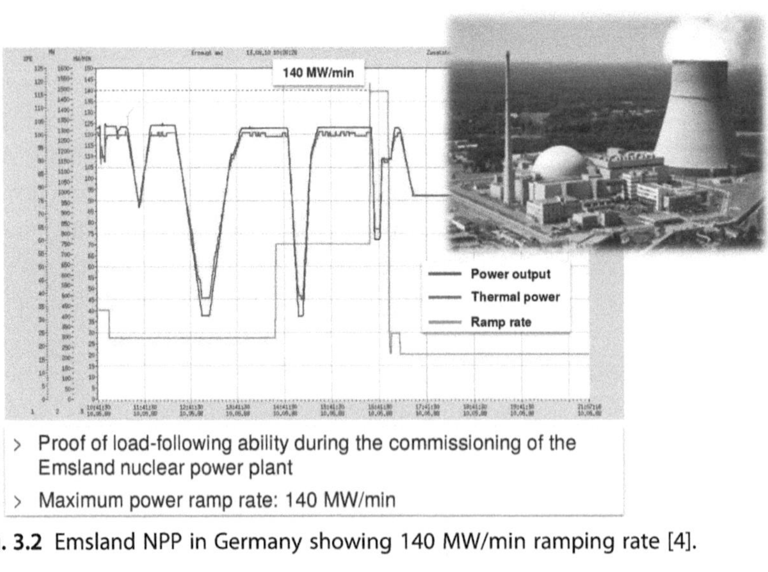

Fig. 3.2 Emsland NPP in Germany showing 140 MW/min ramping rate [4].

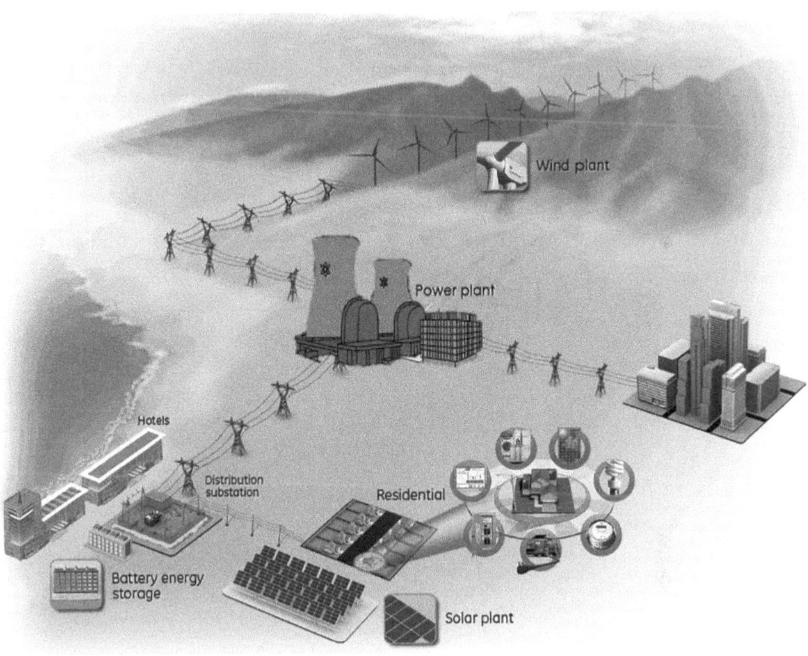

Fig. 3.3 Synergistic existence of nuclear and wind power in a microgrid (10 small wind turbines and 1 NPP—feeding a small locality).

3.2 REDUCED-ORDER MODEL-NUCLEAR REACTOR DYNAMICS

When load demand and in-turn actual reactor power is changed, it has a direct effect on the temperature, pressure, and enthalpy content of the main reactor systems. These thermodynamic changes impact the neutron balance due to changes in the neutron cross-sections of the material constituents. It is indeed a computationally challenging task to capture the detailed 3D thermal hydraulics and neutronics of the reactor core, and due to involvement of different systems the detailed computational assessment is impossible with current resources. Therefore, a reduced-order model is devised to understand the overall reactor system response to obtain temporal evolution of important variables in the lumped fuel or coolant volumes. Point reactor kinetics equations (3.1), (3.2) with effective delayed neutron group are used to model the core neutron population and change in the power levels. The change in thermal power level will impact thermal transport in reactor fuel and coolant systems. A lumped parameter thermal energy transport model is used to assess thermal response. The reactor dynamics and control is captured with the feedback reactivity term that couples the point reactor kinetics and thermal system equations. A reactor will be stable when the reactivity reduces with increase in power level or temperature but it is still critical to understand if the reactor will exhibit a limit cycle due to sustained feedback under all circumstances and what is the expected change in the temperature fluctuations.

$$\frac{dn}{dt} = \frac{(\rho - \beta)n}{\Lambda} + \lambda c \tag{3.1}$$

$$\frac{dc}{dt} = \frac{\beta n}{\Lambda} - \lambda c \tag{3.2}$$

The reactivity feedbacks can be generically defined as

$$\rho = \rho_0 + \rho_c(t) + \rho_f(u_1, u_2, \ldots, u_k) \tag{3.3}$$

$$\frac{du_k}{dt} = f_k(n, u_1, u_2, \ldots, u_k; w_1, w_2, \ldots, w_m) \tag{3.4}$$

3.2.1 Temperature Feedback

Fuel temperature change affects the reactivity primarily due to Doppler broadening and fuel temperature coefficient for reactivity. The resulting impact of the Doppler broadening is to increase the absorption cross-section (e.g., of U-238) resulting into decrease in reactivity. The microscopic fission cross-section will decrease with the temperature leading to additional reduction of the reactivity. Due to these effects, the fuel temperature coefficient for reactivity can be modeled as

$$\alpha_T^F = -\ln\left[\frac{1}{p}\right]\frac{\beta}{2\sqrt{T_F}} \tag{3.5}$$

The coolant or moderator temperature change affects the coolant density, which in turn affects the moderating capacity. Lower moderation of neutrons imply higher resonance absorption, therefore leading to significant reduction in the moderating capacity. Although reduction in coolant density also causes reduction in neutron absorption within the moderator or coolant, the increase in temperature effectively causes negative reactivity in pressurized water reactors (PWRs) and boiling water reactors (BWRs). In BWRs, this effect is due to increase in void fraction and in PWRs it is due to reduction in density of pressurized water. Negative moderator or coolant temperature coefficient is less significant than negative fuel temperature coefficient as changes happen in the fuel immediately. Nevertheless, the overall effect of temperature changes in the fuel and coolant or moderator can be described as Eq. (3.6).

$$\rho = \rho_c(t) + \alpha_T^C \Delta T_C + \alpha_T^F \Delta T_F \tag{3.6}$$

3.2.2 Energy Balance

In the set of equations mentioned previously, Eq. (3.3) expresses the total reactivity in terms of the various contributions. Where ρ_0 is the reactivity under cold conditions with all the control rods withdrawn, $\rho_C(t)$ is the reactivity change owing to various reactivity control mechanisms, and ρ_F is the reactivity introduced into the system by the nonnuclear feedback, for example, temperature changes [5]. The parameter values used for this study are listed in Table 3.1 and parameter definitions are described in nomenclature Table 3.2. These temperature feedbacks can be captured by the differential equations governing the heat balance in the fuel and the coolant regions, respectively [5].

$$m_F C_{pF} \frac{dT_F}{dt} = a_F n - h(T_F - T_C) \tag{3.7}$$

$$m_C C_{pC} \frac{dT_C}{dt} = h(T_F - T_C) - 2W_C C_{pC}(T_C - T_{Cin}) \tag{3.8}$$

Table 3.1 Parameter Values Used in This Reduced-Order Model [5]

β	0.0075
Λ	0.001 s
λ	0.1
C_{pF}	200 J/kg K
C_{pC}	4000 J/kg K
m_F	40,000 kg
m_C	7000 kg
W_C	8000 kg/s
a_F	7×10^6 J/m^3
α_T^F	−0.00001 1/K
α_T^C	−0.00005 1/K
h	4×10^6 W/m^2 K

Table 3.2 Nomenclature Table for the Reduced-Order Model

n	Neutron concentration or density
c	Delayed neutron precursor concentration
β	Delayed neutron fraction
Λ	Average neutron life time
λ	Decay constant
C_{pF}	Specific heat of fuel
C_{pC}	Specific heat of coolant
m_F	Mass of fuel
m_C	Mass of coolant
W_C	Coolant mass flow rate
a_F	Neutrons to thermal energy factor
α_T^F	Doppler feedback coefficient
α_T^C	Moderator feedback coefficient
h	Heat transfer coefficient between fuel and coolant
ζ_{air}	Air density
$\Gamma_{turbine}$	Blade cross-sectional area
$P(t)$	Energy flux
$P_{elec}(t)$	Electrical power
B_p	Betz factor
ϕ_k	Covariance
A_k	Autocorrelation
^{17}O	1/2

where T_C and T_{Cin} in Eq. (3.8) are the coolant average temperature and coolant inlet temperature, respectively.

The model thus described can be written in terms of nondimensional parameters using the relations given in Table 3.3. In which x, y, z_F, z_C are the normalized values of the neutron density, precursor density, fuel temperature, and coolant temperature, respectively. It is to be noted that the variables n_e, C_e, $T_{F,e}$, and $T_{C,e}$ mentioned in Table 3.3 are the equilibrium values of neutron density, precursor density, fuel temperature, and coolant temperature, respectively.

Table 3.3 Nondimensional Variables

x	$\frac{n-n_e}{n_e}$
y	$\frac{C-C_e}{C_e}$
z_F	$\frac{T_F-T_{F,e}}{T_{F,e}}$
z_C	$\frac{T_C-T_{C,e}}{T_{C,e}}$

With these substitutions, the normalized lumped parameter reactor dynamics model becomes,

$$\frac{dx}{dt} = \frac{-\beta x}{\Lambda} + \frac{1}{\Lambda}(\beta y) + \frac{\rho}{\Lambda} + \frac{\rho x}{\Lambda} \tag{3.9}$$

$$\frac{dy}{dt} = \lambda(x - y) \tag{3.10}$$

$$\frac{dz_{\mathrm{F}}}{dt} = \frac{a_{\mathrm{F}} n_{\mathrm{e}} x}{m_{\mathrm{F}} c_{\mathrm{pF}} T_{\mathrm{F,e}}} - \frac{h z_{\mathrm{F}}}{m_{\mathrm{F}} c_{\mathrm{pF}}} + \frac{h T_{\mathrm{C,e}} z_{\mathrm{C}}}{m_{\mathrm{F}} c_{\mathrm{pF}} T_{\mathrm{F,e}}} \tag{3.11}$$

$$\begin{aligned}\frac{dz_{\mathrm{C}}}{dt} &= \frac{h T_{\mathrm{F,e}} z_{\mathrm{F}}}{m_{\mathrm{C}} C_{\mathrm{pC}} T_{\mathrm{C,e}}} - \frac{(2C_{\mathrm{pC}} W_{\mathrm{C,e}} + h) z_{\mathrm{C}}}{m_{\mathrm{C}} C_{\mathrm{pC}}} \\ &\quad + \frac{2W_{\mathrm{C,e}} T_{\mathrm{Cin,e}} u}{m_{\mathrm{C}} T_{\mathrm{C,e}}} - \frac{2W_{\mathrm{C,e}} (T_{\mathrm{C,e}} - T_{\mathrm{Cin,e}}) w}{m_{\mathrm{C}} T_{\mathrm{C,e}}}\end{aligned} \tag{3.12}$$

$$\rho = \rho_{\mathrm{c}}(t) + \alpha_{\mathrm{T}}^{C} T_{\mathrm{C,e}} z_{\mathrm{C}} + \alpha_{\mathrm{T}}^{F} T_{\mathrm{F,e}} z_{\mathrm{F}} \tag{3.13}$$

The variables u and w in Eq. (3.12) are the dimensionless coolant inlet temperature and the dimensionless coolant flow rate, respectively, which are defined in Eqs. (3.14), (3.15) in terms of $T_{\mathrm{Cin,\ e}}$ and $W_{\mathrm{C,\ e}}$, which are the equilibrium coolant inlet temperature and equilibrium coolant flow rate, respectively.

$$u = \frac{T_{\mathrm{Cin}} - T_{\mathrm{Cin,e}}}{T_{\mathrm{Cin,e}}} \tag{3.14}$$

$$w = \frac{W_{\mathrm{C}} - W_{\mathrm{C,e}}}{W_{\mathrm{C,e}}} \tag{3.15}$$

3.2.3 Control System

To simulate the scenario in which the demand side is dictating the system output, it is necessary to introduce an appropriate controller in the system so that the system may adapt to the changing demand. A proportional controller is used in this analysis, and excess reactivity is introduced via a classical output feedback control. In other words, a unity feedback control loop shown in Fig. 3.4 is implemented. Based on the mismatch between the power being produced and power demand, externally controlled reactivity is varied in such a way so as to minimize the mismatch.

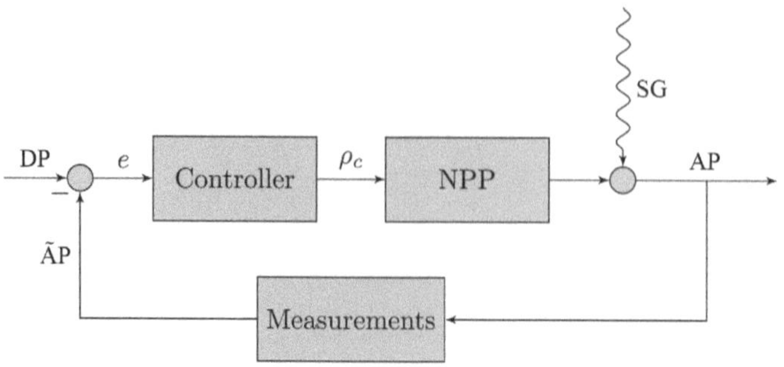

Fig. 3.4 Classical output feedback control. *AP*, actual power; *DP*, demand power; *SG*, stochastic generation.

The reduced-order model equations along with the proportional control action were numerically solved using the Runge-Kutta-Fehlberg (RK45) method [6, 7]. Time step for these calculations was 0.001 s, and total simulation time was 150 s. The two scenarios are considered to understand the reactor behavior—(a) in first scenario, the demand power level is increased by 13% for 2 min and then brought back to the same level. This scenario approximately depicts the controlled load following reactor operations for some European NPPs described earlier in the introduction. (b) The second scenario illustrates the sudden changes in demand power as the wind energy plant and NPP compliment each other to supply the electrical load for a microgrid. The wind energy generation is highly stochastic in nature; therefore, a stochastic and realistic model is developed based on wind speed time series data to compute the demand power estimates. The methodology to construct energy generation from wind power is described next, which is used to obtain controller signal.

3.3 ESTIMATING POWER GENERATION FROM WIND ENERGY SYSTEMS

For a given time-dependent wind speed data, the instantaneous electrical energy production rate at the outlet of the wind turbine can be calculated with the following approach. Considering "10" small wind turbines, each of blade radius 12 m, the kinetic energy flux of the volumetric air can be written as:

$$E(t) = \frac{1}{2}\zeta_{\text{air}}(\| v(t) \|)^2 \tag{3.16}$$

and the corresponding energy flux can be written as

$$P(t) = E(t) \times v(t) \tag{3.17}$$

It is important to understand that energy flux is the total power available in the air for extraction—which of course is limited by the design factors of the wind turbine. To convert the net available power to the net electrical power, Eq. (3.17) should be multiplied by the cross-sectional area of the wind turbine blades (Γ_{turbine}) and by the efficiency factor or the power coefficient (B_{p}) of the wind turbine. This factor (B_{p}) is also known as "Betz" factor, which has an upper limit of 59%. Assuming a "Betz" factor of 0.3 or a 30% efficient turbine:

$$P_{\text{elec}}(t) = P(t) \times 0.3 \times \Gamma_{\text{turbine}} \tag{3.18}$$

3.3.1 Statistical Model of Wind Speed

To calculate the kinetic energy of the wind (Eq. 3.16) the raw velocity data can be fitted into a statistical model. From the perspective of the economic dispatch of electrical power and the operation of the power systems, it is required to predict the wind speed on a shorter time scales as compared to reported data [8, 9]. Typically for that reason an AR-1 model is used which assumes that the wind time series with some known statistical parameters is totally random and there are no memory effects involved. The AR-1 model can be written as:

$$\tilde{Y}_t = \gamma \tilde{Y}_{t-1} + R(t) \tag{3.19}$$

The previous model is also known as Markov process, which can be attributed to the fact that the current value of the variable $\tilde{Y}$ only depends on the value immediately behind it in time.

The process described in Eq. (3.19) is said to be stationary in a sense that its properties are unaffected by change of time origin, that is the joint distribution of the n observations made at any times $(\tilde{Y}_{t_1}, \tilde{Y}_{t_2}, \ldots, \tilde{Y}_{t_n})$ is the same as that associated with n observations made at any times $(\tilde{Y}_{t_1+k}, \tilde{Y}_{t_2+k}, \ldots, \tilde{Y}_{t_n+k})$. In other words the joint distribution of variables will not get affected by shifting all the times of observation forward or backward by an integer amount k [10]. Within this it can be observed that for $n = 1$ the stationarity criteria imply that the probability density of the process $\tilde{Y}_t$ is constant for all times.

A simple AR-1 stationary process is used with a constant value of γ such that $|\gamma| < 1$ and with a constant variance of $R(t)$. The $R(t)$ term is white noise, that is, it is uncorrelated in time and is centered on mean "0" [10].

3.3.2 Load Curve for NPP

The k time step autocorrelation A_k is "0.8" and k value is set to the number of time steps in an hour $k = \frac{60}{t}$ where t is the time step. With this the γ can be calculated from Eq. (3.20).

$$A_k = \gamma^k \tag{3.20}$$

Once all the values of AR-1 model were calculated, a standard Monte-Carlo program was used to obtain model realizations. These realizations of the synthetically generated time series are normally distributed but the real wind time series is Weibull distributed. Thus, the probability distribution function of the generated series should be transformed. This can be achieved by the following process:

$$\frac{f(\tilde{u}) = p(\tilde{\gamma})d\tilde{\gamma}}{du} \tag{3.21}$$

$$\int_a^b f(\tilde{u})d\tilde{u} = \tilde{\gamma}, \quad \forall \tilde{\gamma} \in \text{uniform distribution} \tag{3.22}$$

$$\text{if } f(\tilde{u}) \text{ is Weibull distributed}, \quad \tilde{\gamma} = 1 - e^{(-\frac{\tilde{u}}{\lambda})^p} \tag{3.23}$$

$$\text{If } \tilde{\gamma} \in \mathcal{N} \text{ and } \tilde{u} \text{ is uniform}, \quad \tilde{u} = 0.5\left(1 + \text{erf}\left(\frac{\tilde{\gamma} - \mu}{\sigma\sqrt{2}}\right)\right) \tag{3.24}$$

Using the earlier two results one can get $\tilde{u} = \lambda \frac{(-\log((1-\text{erf}\tilde{\gamma} - \mu}{\sigma\sqrt{2}})))^{1/p}$ [11, 12], to convert a normally distributed variable $\tilde{\gamma}$ to a Weibull distributed variable (Fig. 3.5).

The power output was predicted from the wind speed distributions and then used in the microgrid model to obtain instantaneous total demand power (DP) from the reactor using total power (TP) and wind power output

$$\text{DP} = \text{TP} - p(t) \tag{3.25}$$

The predicted instantaneous demand power is used to predict reactor behavior using reduced-order model and controller system described in previous section.

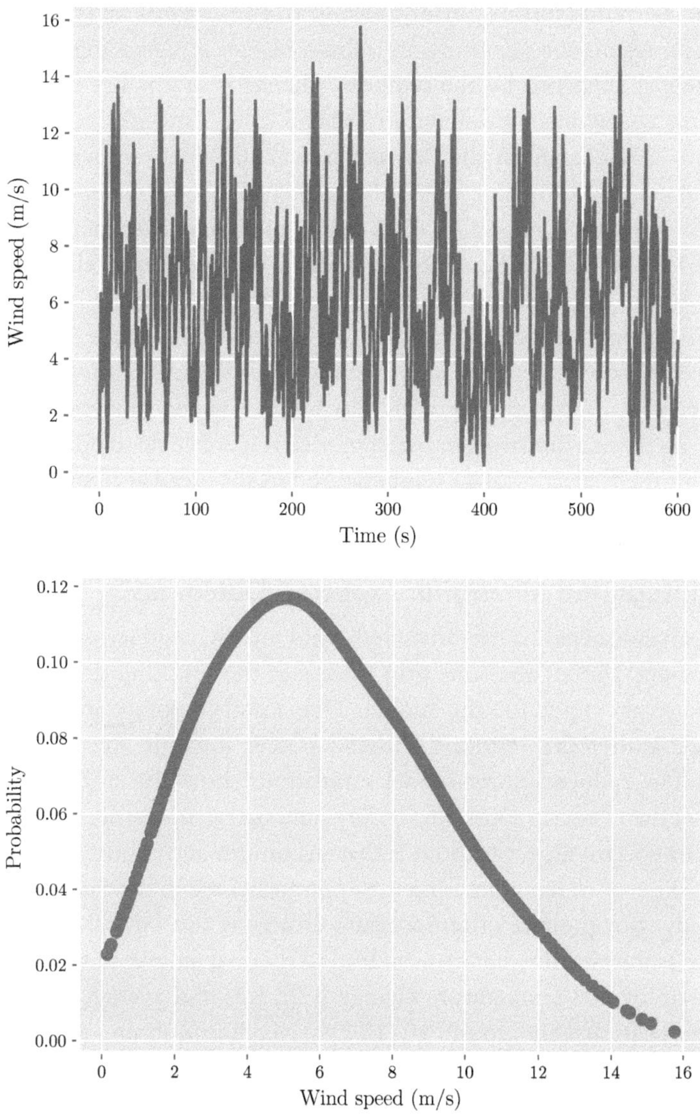

Fig. 3.5 Generated wind speed time series and the probability distribution.

3.4 NUMERICAL RESULTS

First, example test for this load following response involves the power demand increase 13% for a fixed time period of 120 seconds and then followed by reduction in power demand to the original level. Numerical simulation results for this study are shown in Fig. 3.6, and the response of the system is observed by the temporal change in normalized variables x, y, z_F, and z_C, which are defined in Table 3.2.

For a 13% change in the demand power, the fuel temperature jumps from the equilibrium value of 921.88 to 967.97 K over a period of 120 s, that is, a relative change of 46.09 K while the coolant temperature jumps from 571.88 to 574.73 K. Thus the coolant experiences a relatively small change in the temperature as compared to the fuel. This is because fluid is flowing at high mass flow rate and has an additional convection term as compared to solid stationary fuel. Even with small change in coolant temperature, it is important to examine its impact on the subcooled nucleate boiling. As this is a lumped parameter model, it is only able to predict average change in the fuel or coolant temperature but for more accurate analysis at least 1D models are required to capture the effects of axial or radial peaking.

3.4.1 Integrated Microgrid: Expected Outcomes

In the second scenario, the predicted wind energy production rate (maximum power 13% of the total grid load) and the resulting demand power are used as an input for the model. Due to stochasticity in wind speed described earlier the resulting demand power input is also stochastic in nature. The reduced-order model simulations show the stochastic effects appear in the resulting solutions for fluid and fuel temperatures. One significantly important aspect to note is that maximum and minimum temperatures of the system remain within the limits presented in the results from first case study. But the rate of temperature change is also critical parameter to estimate as thermal fluctuations can lead to thermal fatigue or failures. Maximum rate of fuel temperature change is 22 K/s and average temperature fluctuations in fuel are around 10–15 K/s. Maximum strain amplitude during these thermal fluctuations is expected to be 0.01%. If the frequency of these fluctuations is much less than the natural frequency of the PWR fuel cladding (i.e., 50 Hz), these thermal effects are not expected to cause any major concern to the fuel integrity (Fig. 3.7).

3.5 SUMMARY

A reduced-order model of the pressurized water reactor for the simulation of the load following operations is assessed and tested with a

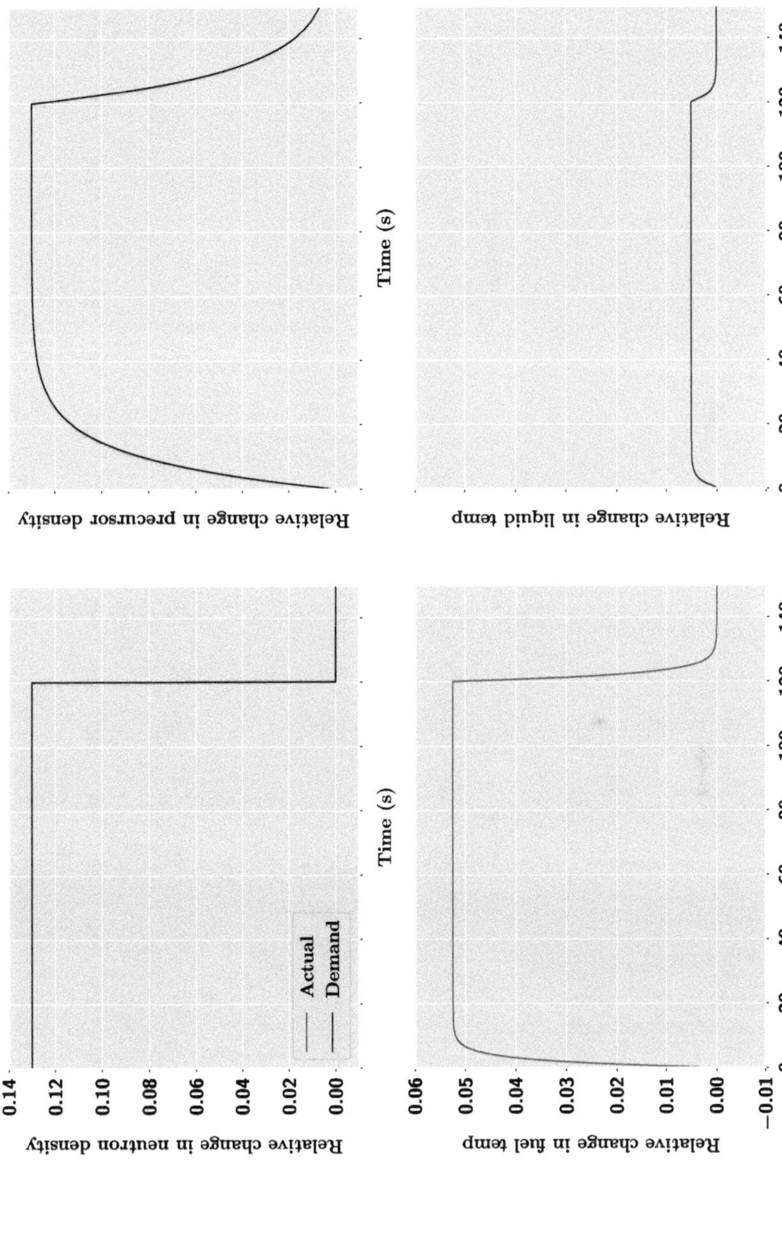

Fig. 3.6 Response of the reactor dynamics model (x, y, z_F, z_C from left to right). Time scale is in "seconds."

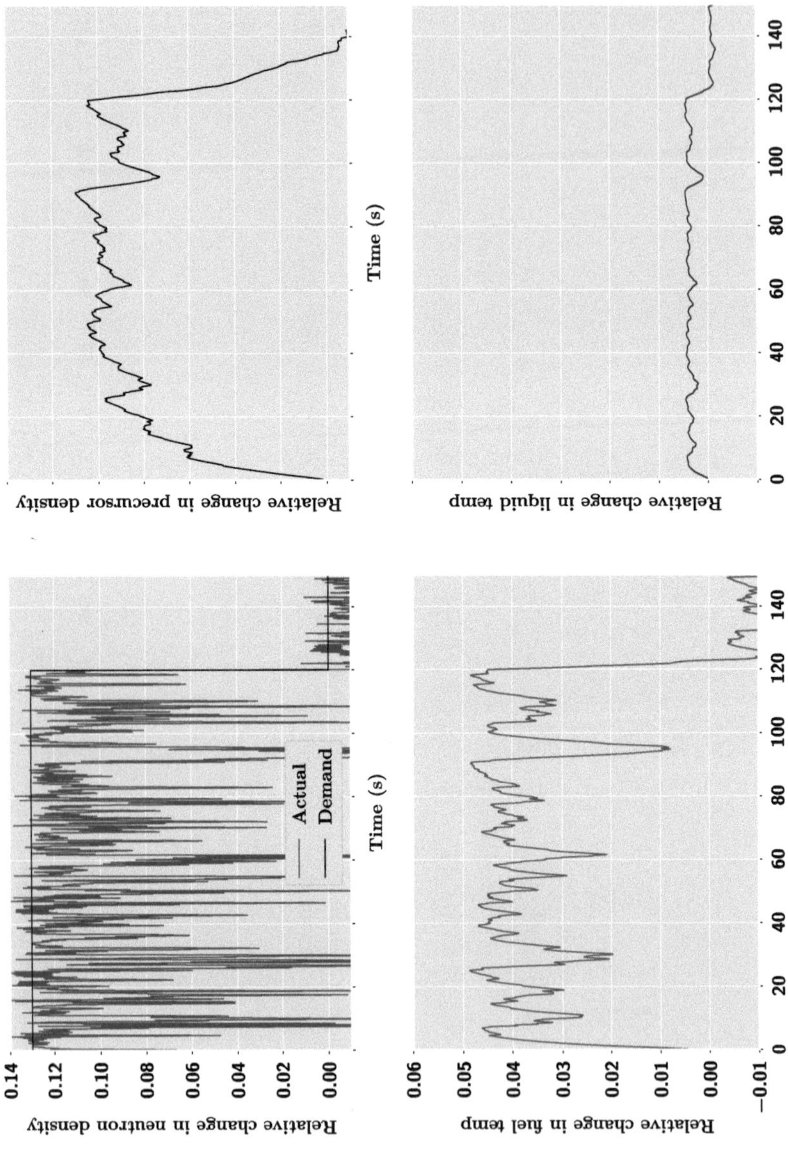

Fig. 3.7 Response of the reduced-order model owing to stochastic and deterministic generation.

proportional controller in the classical feedback control loop. In the first example, model was used to simulate the variations in the fuel and coolant temperature following a step change in demand power governed by the grid. The response of the reactor system to this 120-s transient shows that the system is capable to achieve new steady state quickly. Change in coolant temperature is anticipated to be much smaller than change in the fuel temperature.

Wind and nuclear energy powered microgrid is studied with this reduced-order model. The realistic wind speed distribution is used to construct autoregressive model and predict instantaneous power output from wind energy system. Deficit power in the microgrid is supplied by nuclear reactor, thus stochastic demand power input is used to predict reactor thermal behavior. Thermal response of fuel and the coolant shows that temperature limits do not exceed the previous case with maximum power of wind energy system within 13% of the total grid load. Under these conditions, thermal fluctuations are monitored closely to assess thermal fatigue problems. The maximum strain amplitude is with in the failure limits of the nuclear fuel cladding but it is important to ensure that frequency of thermal cycles remain below the natural frequency of the cladding. Although the model results show stable reactor behavior for controlled load-following or small fraction of highly variable load demand for more accurate analysis particularly related to thermal fluctuations, it is important to improve the fidelity of the models.

In future it is advisable to construct a reduced-order model by more rigorous techniques such as proper orthogonal decomposition and/or singular value decomposition—("Law of Parsimony") [13]. The wind speed model considered here is a point estimate of the wind speed data while a more realistic estimate can be more appropriately given by the space time correlated models. For a large wind generation component, geographical assessment should be considered. Geographical smoothing/coupled behavior of two geographically isolated renewable power plants is inherently intertwined to the grid scale power fluctuations—which may not (geographically smoothened) or may affect (coupled) the conventional power generators like nuclear [14].

REFERENCES

[1] NERC, Accommodating High Levels of Variable Generation, Intermittent, NERC and Force, Variable Generation Task, North American Electric Reliability Corp., 2009.

[2] F. Huntowski, A. Patterson, M. Schnitzer, Negative Electricity Prices and the Production Tax Credit, The North Bridge Group, Concrod, MA, 2012.
[3] A. Lokhov, Technical and Economic Aspects of Load Following With Nuclear Power Plants, NEA, OECD, Paris, France, 2011.
[4] C. Bruynooghe, A. Eriksson, G. Fulli, Load-following operating mode at nuclear power plants (NPPs) and incidence on Operation and Maintenance (O&M) costs, in: Compatibility With Wind Power Variability, 2010.
[5] H. Anglart, Nuclear Reactor Dynamics and Stability, KTH Royal Institute of Technology, Sweden, 2011.
[6] E. Fehlberg, Low-order classical Runge-Kutta formulas with stepsize control and their application to some heat transfer problems, (1969), NASA Technical Report, NASA TR-315.
[7] E. Hairer, W. Gerhard, Solving ordinary differential equations ii: stiff and differential-algebraic problems second revised edition with 137 figures, in: Springer Series in Computational Mathematics, vol. 14, 1996.
[8] X. Zhu, M.G. Genton, Short-term wind speed forecasting for power system operations, Int. Stat. Rev. 80 (1) (2012) 2–23.
[9] P. Pinson, Wind energy: forecasting challenges for its operational management, Stat. Sci. 28 (4) (2013) 564–585.
[10] G.E.P. Box, G.M. Jenkins, G.C. Reinsel, G.M. Ljung, Time Series Analysis: Forecasting and Control, John Wiley & Sons, Hoboken, NJ, 2015.
[11] D. Villanueva, A. Feijóo, J.L. Pazos, Multivariate Weibull distribution for wind speed and wind power behavior assessment, Resources 2 (3) (2013) 370–384.
[12] D. Villanueva, J.L. Pazos, A. Feijoo, Probabilistic load flow including wind power generation, IEEE Trans. Power Syst. 26 (3) (2011) 1659–1667.
[13] A.C. Antoulas, Approximation of Large-Scale Dynamical Systems, SIAM, Philadelphia, PA, 2005.
[14] M.M. Bandi, Spectrum of wind power fluctuations, Phys. Rev. Lett. 118 (2017) 028301.

CHAPTER FOUR

Nuclear Hydrogen Production

Shripad T. Revankar
School of Nuclear Engineering, Purdue University, West Lafayette, IN, United States

4.1 INTRODUCTION

4.1.1 Need for Hydrogen Production

Fossil fuels in the form of coal, oil, and natural gas have been used to power industrial technology and transportation since the 18th-century industrial revolution. The demand for fossil fuel has increased over the years, and the world's demand for energy is projected to double by 2050 with expected population growth and the industrialization of developing countries. There are major issues with the continued use of fossil fuels. The supply of the fossil fuel is limited, and their reserves are concentrated in a few regions of the world. A secure supply of fossil fuel with increased demand worldwide is increasingly difficult to assure. Moreover, the fossil fuels being hydrocarbons when used emit particulate pollutions, carbon dioxide, and other greenhouse gases. The pollutions increase human health risk, and the greenhouse gases cause global warming threatening the stability of Earth's climate. In this context, hydrogen has been found to be a potential sustainable clean energy carrier with low-cost method of producing it in large capacities to replace fossil fuel. But hydrogen, though the most common chemical element on the planet, does not occur in nature as the fuel. Like electricity, hydrogen is an energy carrier, and it has to be separated from chemical compounds, by electrolysis from water or by chemical processes from hydrocarbons or other hydrogen carriers. Most of the world's hydrogen (about 97%) is currently derived from fossil fuels through some type of reforming process such as steam-methane reforming (SMR). Hydrogen is used widely by petrochemical, agricultural (e.g., ammonia for fertilizers), manufacturing, food processing, electronics, plastics, metallurgical, aerospace, and other industries. The economic infrastructure based on hydrogen energy carrier is called hydrogen economy, which is composed of three functional steps, production, storage, and transportation, and used in all aspects of the economy. Fig. 4.1 shows the essential features of hydrogen economy.

Storage and Hybridization of Nuclear Energy
https://doi.org/10.1016/B978-0-12-813975-2.00004-1

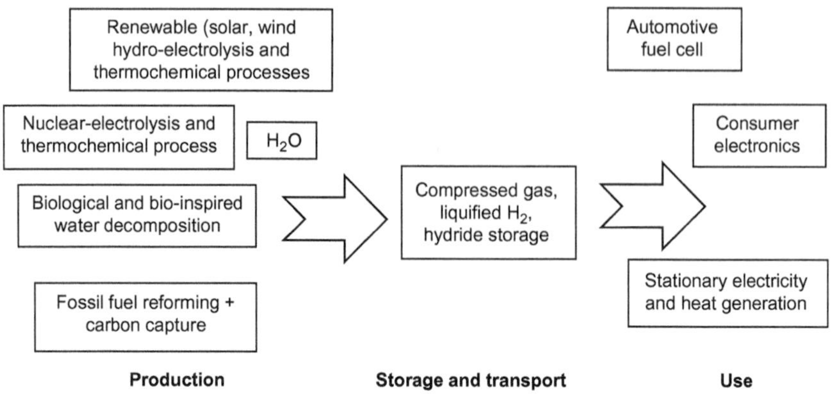

Fig. 4.1 Hydrogen economy.

Hydrogen can be produced in number of ways using any of the primary energy sources. Of the four primary sources, fossil, geothermal, nuclear, and solar (renewable), nuclear and solar are the practical energy sources with no greenhouse gas emission. For large-scale hydrogen production, nuclear energy is well suited. Developing a hydrogen economy is a growing national and international issue as hydrogen demand continues to rise and cleaner energy sources and carriers are desired. As per 2014 data, more than 58 million tons are produced and consumed globally each year. A very small amount of this hydrogen is used as a fuel; most of it is used for commercial chemical use such as for fertilizer, petroleum refining, hydrocracking, hydrodesulphurization, welding, metallic ore reduction, hydrochloric acid production, and in space propulsion with liquid hydrogen. There is a large demand for the hydrogen economy, and it is increasing at nearly 4%–10% a year. A big advantage to producing hydrogen from nuclear energy as opposed to other energy forms is that one nuclear pellet (1 cm diameter) could feasibly make enough hydrogen to fuel 220 cars for 100 miles. Producing hydrogen on a large scale involves the coupling of an energy production method to a process or system of processes that will allow for hydrogen production. Although hydrogen has been commercially produced for decades, the utilization of nuclear power in conjunction with a suitable method of producing hydrogen is a sustainable technology without CO_2 production.

4.1.2 Hydrogen

Hydrogen is the first and lightest element in the periodic table and the most abundant element in the universe constituting 75% of mass or 90% of volume. On Earth, it is mostly found in molecular compounds such as water and organic matters as it readily forms covalent compounds with most nonmetallic elements. Compounds of hydrogen are also found in the form of hydrides, such as LiH and BeH, where hydrogen atom has acquired a negative or anionic character. It also exits as a free element in the atmosphere in a very small amount, <1 ppm by volume. Henry Cavendish was first to identify hydrogen as a distinct species in 1776. Seven years later, Antoine Lavoisier named it, in Greek, hydrogen, which means "water maker," and showed that water was composed of hydrogen and oxygen. Hydrogen has three naturally occurring isotopes: stable protium or hydrogen (1H or H) with natural abundance of over 99.98%; stable deuterium (2H or D) or heavy hydrogen with natural abundance of 0.014%, discovered by Harold Urey in 1932; and radioactive tritium (3H or T) with half-life of 12.3 years, discovered by Ernest Rutherford in 1934.

Gas hydrogen is a colorless, odorless, tasteless, nontoxic, noncorrosive, and nonmetallic diatomic gas. Hydrogen behaves like an ideal gas over modest range of temperature and pressures (<5 MPa) where the density ρ can be obtained by using ideal gas law $\rho = P/RT$, with the specific gas constant $R = 4124.45$ J/kgK. However, at high pressures, it shows significant deviation from ideal gas law, and the density of hydrogen at 20°C is actually 2.9% less at 5 MPa and 5.7% less at 10 MPa than predicted by the ideal gas law. At standard temperature and pressure (STP) conditions (20°C and 101.325 kPa), density of hydrogen gas (H_2) is 0.08375 kg/m^3, and specific volume is 11.940 m^3/kg. The density of hydrogen is 23.651 kg/m^3, and specific volume is 0.042282 m^3/kg at 35 MPa, while these values are 39.693 kg/m^3 and 0.025193 m/kg at 70 MPa, respectively. The other physical properties of hydrogen are given in Table 4.1.

Hydrogen gas is highly diffusive and highly buoyant; it rapidly mixes with the ambient air upon release and can diffuse through materials that are impermeable to other gases. Hydrogen is generally not toxic but poses a risk of asphyxiation if inhaled. Because hydrogen gas has no color, odor, or taste, it is difficult to detect a leak from a tank. Hydrogen can easily mix with ambient air, and the rapid mixing of hydrogen with the air is a safety concern, since it can lead to flammable mixtures. When mixed with oxygen, hydrogen is highly flammable over a wide range of concentrations. The specific energy density of hydrogen is very high; 1 kg of hydrogen contains 141.86 MJ, which is approximately 2.5 times more energy than is contained in 1 kg of natural gas. The energy content of hydrogen is given as either LHV of 242 kJ/mol or as HHV of 286 kJ/mol. A stoichiometric hydrogen-air mixture, where all fuel is consumed upon reaction, contains 29.5 vol% of hydrogen with the combustion product as water vapor. Hydrogen flame

Table 4.1 Hydrogen Properties

Phase at STP	Gas
Standard atomic mass	1.0078225 u
Protium	1.00782503207 u
Deuterium	2.0141017778 u
Tritium	3.0160492777 u
Molecular weight	2.01594 g/mol
Melting point	13.99 K (−259.16°C or −434.49°F)
Boiling point	20.271 K (−252.879°C or −423.182°F)
Density (at STP)	0.08988 g/L
Density when liquid (at m.p.)	0.07 g/cm^3 (solid, 0.0763 g/cm^3)
Density when liquid (at b.p.)	0.07099 g/cm^3
Triple point	13.8033 K, 7.041 kPa
Critical point	32.938 K, 1.2858 MPa, 31.263 kg/m^3
Heat of fusion	0.117 kJ/mol
Heat of vaporization	0.904 kJ/mol
Molar heat capacity	28.836 J/(mol K)
Speed of sound	1310 m/s (gas, 27°C)
Thermal conductivity	0.1805 W/(m K)
Expansion ratio liquid/ambient	845
Diffusion coefficient at NTP	0.61×10^{-4} m^2/s
Ratio of specific heats cp/cv at NTP	1.308
Viscosity gas at NTP	8.948 μPa s
Surface tension at BP	1.92×10^{-3} N/m
Speed of sound in gas at NTP	1294 m/s
Hot air jet ignition temperature	943 K
Autoignition temperature in air	793–1023 K
Minimum ignition energy for air H_2 mixture	0.017 mJ
Minimum ignition energy	1.9×10^{-5} J
Minimum ignition energy for detonation	~10,000 J
Maximum flame speed for stoichiometric H_2-air mixture	404 m/s
Flammability limits in air	4.0–75.0 vol%
Detonatability limits in air	13–70 vol%
Inversion (Joule-Thomson) temperature	193 K
Adiabatic flame temperature	2318 K
Visible laminar flame front velocity	18.6 m/s
Detonation velocity	1480–2150 m/s
Energy release	2.82 MJ/kg mixture

1 u (atomic mass unit) = $1.660538782 \times 10^{-27}$ kg.
Normal temperature and pressure (NTP): 293 K and 101,325 Pa.
Standard temperature and pressure (STP): 273 K and 101,325 Pa.

is a nonluminous, almost invisible pale blue with the flame temperature of 2403 K. The flammability range for hydrogen (at room temperature) ranges from 4 and 75 vol% in air to 95 vol% in oxygen. The flammability range widens with higher temperatures. The potential for an explosion of a flammable hydrogen-air mixture is very high. The autoignition temperature, which is the minimum temperature of a hot surface that can ignite a flammable mixture, is for hydrogen in the range of 800–1000 K dependent on the experimental conditions. The minimum ignition energy, to ignite the most easily ignitable hydrogen concentration in air, is very low (0.017 MJ), much lower than for hydrocarbon-air mixtures. The detonatability range of hydrogen is usually given to be 13–70 vol% of hydrogen concentration.

Hydrogen reacts both with nonmetals and metals to form either ionic or covalent hydrides (e.g., HCl and H_2O). In the measure for the attraction of electrons to the nucleus, the electronegativity of hydrogen is 2.20 on the Pauling scale. Hydrogen forms host of compounds known as hydrocarbons when reacted with carbon, including natural gas and crude petroleum. Hydrogen highly reactive property is used in a broad range of manufacturing activities such as in sweetening organic components in petroleum (refineries) and food processes and ammonia fertilizers through chemical reaction of hydrogen with the nitrogen gas in the air. It is used as an effective reducer to remove oxygen (forming H_2O) from metal oxides to produce metals such as in the steel industry. It is also used to chemically remove unwanted impurities from industrial products.

Globally, over 58 million tons of hydrogen was consumed in 2014 each year for use as a critical feedstock for the production of clean burning transportation fuels, fertilizers, and host of chemicals (BCC Research, "Building the Global Hydrogen Economy: Technologies and Opportunities," November 2015). The demand for hydrogen is expected to increase by 5% per year until 2023.

Hydrogen is not found free in nature and must be derived from four main primary energy sources: fossil energy, renewable energy, geothermal energy, and nuclear energy. Currently, most hydrogen is produced from fossil energy using steam-methane reforming (SMR) of natural gas, followed by partial oxidation (POX) and autothermal reforming (ATR), which combines SMR and POX processes. The use of fossil fuel for hydrogen production produces carbon dioxide, a greenhouse gas that has been attributed to global warming and pollution. Hydrogen is proposed as a solution to this environmental issue through a concept called hydrogen energy economy. Under hydrogen economy, hydrogen is produced economically and environmental friendly, stored and carried as industrial material, and of sufficient quantities to replace fossil resources (oil, natural gas, and coal) that are used in today's fossil energy economy. This concept is being developed because the current practice of fossil energy economy is considered to be unsustainable and leads to build of greenhouse gas. Under this economy, hydrogen can be produced from non-CO_2 emitting primary sources such as nonfossil energy sources, solar, ocean thermal, wind, waves, thermonuclear, geothermal, etc., and can be used in a full range of applications in all sectors of the economy: transportation, power, industry, and buildings. Given intermittent nature of renewable energy requiring large energy storage currently it is limited by economic viability for large-scale production of hydrogen with renewable energy. On the other hand, the nuclear energy can provide for large-scale hydrogen production capability. Nuclear energy produces primarily heat source, which is then converted to electric energy through Rankine or Brayton cycle. Other than hydrocarbons, the major source of hydrogen is water, which is quite abundant. Hydrogen can be produced by decomposing water, that is, breaking the chemical bonds of water. The decomposition of water can be facilitated by electrolysis, direct thermal decomposition, chemical reaction, and thermochemical cycles.

4.1.3 Nuclear Energy

Nuclear energy comes either from two nuclear reactions, fission and fusion. In the fission reaction, an atom with large atomic number such as thorium (232), uranium (235), or plutonium (239) is split into two or three lower atomic number elements when a neutron is absorbed. The fission releases large quantity of energy. The fusion is exactly an opposite process where two or more of the atoms of the hydrogen and its isotopes, deuterium and tritium, fuse into a heavier element accompanied by the release of large amount of energy. The energy released in fission or fusion is a result of loss of mass (m) in the reaction products, and this energy (E) is given by Albert Einstein in 1905, with his famous equation $E = mc^2$, where c is the speed of light. The atoms that release this atomic energy are feed material or fuel for the nuclear reactors, and these are naturally abundant and also can be synthesized from other abundant sources. The fission reactor technology is highly matured technology and currently around 450 commercial nuclear reactors are operating around the world (April 2017) and some 60 more are being constructed.

In the fission nuclear reactors, the nucleus of fuel absorbs a striking neutron and undergoes fission producing more than one neutron that are then absorbed by other nuclei of the same isotope to produce further fission and even more free neutrons. When this process repeats continuously, the result is a chain reaction. Such isotopes that enable a chain fission reaction are termed fissile materials and are directly useable as nuclear fuel in a nuclear reactor. A typical fission reaction in ^{235}U is shown in Fig. 4.2. Uranium-235 can absorb the low-energy neutrons (~0.1 eV) called thermal neutrons that approach uranium nucleus at low speeds (on the order of 104 m/s). The neutrons are born in fission with high energy (~2 MeV) and are fast at about 7% of the speed of light. These so-called fast neutrons are too energetic to be caught by ^{235}U, and fission becomes improbable, and in nuclear reactors, a moderator, which can be water or graphite, is used to slow the fast neutrons down to thermal neutrons to induce fission. The uranium isotope uranium-238 (^{238}U), which is 99.35% naturally abundant of uranium (the balance 0.65% is ^{235}U), is not fissile. But it is able to absorb a fast neutron and breed the transuranic isotope of ^{239}Pu, which is fissile. For this reason, ^{238}U is said to be fertile. Similarly, thorium-232 is fertile, which can absorb a thermal neutron to breed the fissile ^{233}U.

Fission reactions vary greatly with number of fission product generated including two new nuclei with atomic masses close to 95 and 135 u, two or three neutrons, several γ-ray photons, and large release of energy. The fission fragments from the fission of ^{235}U include hundreds of pairs of elements such as Kr and Ba, Zr and Te, and Cs and Rb. Two specific reactions of ^{235}U fission are shown below:

$$^{235}_{92}\mathrm{U} + {}^{1}_{0}\mathrm{n} \rightarrow 2\,{}^{1}_{0}\mathrm{n} + {}^{100}_{38}\mathrm{Sr} + {}^{134}_{54}\mathrm{Xe} + \text{energy}\ (\sim 212\,\mathrm{MeV})$$
$$^{235}_{92}\mathrm{U} + {}^{1}_{0}\mathrm{n} \rightarrow 3\,{}^{1}_{0}\mathrm{n} + {}^{89}_{36}\mathrm{Kr} + {}^{144}_{56}\mathrm{Ba} + \text{energy}\ (\sim 212\,\mathrm{MeV})$$

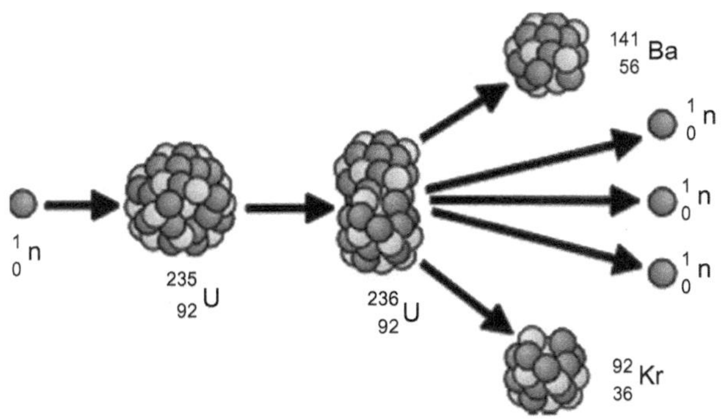

Fig. 4.2 Fission in ^{235}U with daughter nuclei ^{141}Ba and ^{92}Kr and three neutron emissions.

The total energy released from ^{235}U fission is approximately 212 MeV, of which 80% of the energy about 169 MeV is associated with the pair of fragments in the form of kinetic energy that fly apart at about 3% of the speed of light due to Coulomb repulsion and emit about 7 MeV γ-ray photons from recoil. The neutrons that are emitted at the instant of fission called prompt neutrons and the delayed neutrons emitted after the fission carry a sum of about 14 MeV. The total prompt fission energy amounts to about 190 MeV or ~90% of the fission energy. The remaining ~10% of energy is released from β^- decays of the fission fragments to daughter isotopes, delayed neutrons and γ-rays from the decays, and antineutrinos. In a nuclear reactor with the exception of the energy associated with antineutrinos, all other energy is recovered as heat. The total useable energy from fission is a little over 200 MeV or 32 pJ ($1\,MeV = 1.602 \times 10^{-13}\,J$) for any of the three fissile materials ^{233}U, ^{235}U, and ^{239}Pu. A gram of ^{235}U contains 2.56×10^{21} atoms, and with same number fission reactions, 84.1 billion Joules energy is released. This is equivalent to the energy released by combusting 3.5 tons of coal, 2200 L of oil, or 2100 m^3 of natural gas. Fission of a kilogram of ^{235}U can generate electricity for use by half a million homes for a day. In a reactor, the fuel is usually in the form of uranium oxide UO_2 with fissile uranium enrichment of a few percent (1%–5%), and <20% is loaded in a fuel assembly. Several fuel assemblies are arranged to form a core of critical assembly needed to enable a chain fission reaction.

The reactor power is controlled by control rods containing material of cadmium, hafnium, or boron that are effective neutron absorbers. They are inserted into or withdrawn from the core to decrease or increase the population of free neutrons that cause fission. Reactor heat is removed by fluid called coolant including water, gas, liquid metal, helium, carbon dioxide, or molten fluoride salt. The heat carried away from the core is transported to the balance of a reactor plant for the conversion of heat to electricity through Rankine or Brayton cycle. This heat and/or electricity can be used to produce hydrogen. The coolant is looped back into the core to continue the cyclic process of fission-heat production, transfer, and transport.

4.2 NUCLEAR-BASED HYDROGEN PRODUCTION

As nuclear energy is a long-term energy resource that can serve the world for centuries without the emission of any CO_2 or any toxic air pollutants such as those emitted by fossil-fueled power plants, it is the best primary energy source for large-scale hydrogen generation with water splitting by electrolysis or by thermochemical processes. Current operating light-water reactors operate in the range of 280–330°C and primarily generate electrical power. Though this temperature is low for hydrogen production using steam-methane reforming (SMR) or other thermochemical process; current nuclear reactors can be used to generate hydrogen using electrolysis of water. More efficient hydrogen production can be attained by thermochemical splitting of water or electrolysis of high-temperature steam. Efficient water-splitting processes and nuclear SMR require temperatures well above 700°C. Advanced reactors, such as gas-cooled reactors, can achieve the required high temperatures. Here, various nuclear-based hydrogen generation technologies are discussed.

4.2.1 Low-Temperature Water Electrolysis

Electrolysis has been in use since the 1890s to split water into hydrogen and oxygen to supply primarily for chemical industries. Though electrolysis is more expensive than steam reforming of natural gas, it has advantages for generating hydrogen at small facilities for localized high-purity hydrogen and oxygen supplies, and it can meet the need of storage requirement for intermittent renewable technologies. Electrolysis technologies can be divided in two basic categories, alkaline liquid electrolyte using potassium hydroxide (KOH) and acid electrolyte with solid polymer as a proton-exchange membrane (PEM). In both technologies, water is fed into the reaction electrolyte and is subjected to an electric current that causes dissociation, after which the resulting hydrogen and oxygen atoms are put through an ionic transfer mechanism that causes the hydrogen and oxygen to accumulate in separate physical streams. Electrolysis of water is a combination of two half reactions as shown below for acid and alkaline electrolytes:

Acid electrolyte:

$$\text{Anode}: H_2O \rightarrow 2H^+ + 0.5O_2 + 2e^-$$
$$\text{Cathode}: 2H^+ + 2e^- \rightarrow H_2$$

Alkaline electrolyte:

$$\text{Anode}: 2OH^- \rightarrow H_2O + 0.5O_2 + 2e^-$$
$$\text{Cathode}: 2H_2O + 2e^- \rightarrow H_2 + 2OH^-$$

Overall reaction:

$$H_2O \rightarrow H_2 + 0.5O_2$$

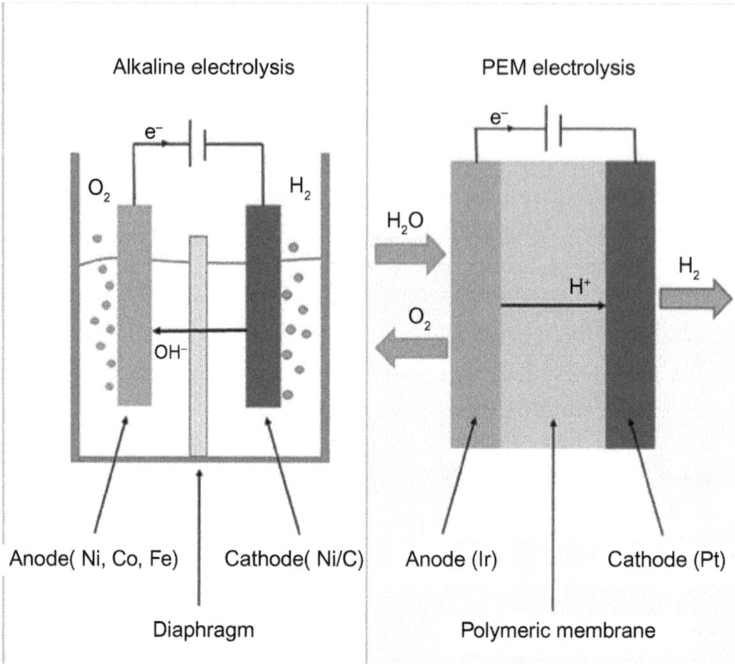

Fig. 4.3 Operation principles of alkaline and proton-exchange membrane (PEM) water electrolysis.

Fig. 4.3 shows the operation principle of electrolysis with alkaline and PEM electrolyzers. Oxygen evolution occurs at the anode; hydrogen evolves at the cathode. In an alkaline electrolysis cell containing an aqueous caustic solution with usually 20%–40% KOH or NaOH (typically 30%), electric energy is applied to two electrodes. The cathode electrodes typically are made of low-carbon steel mesh or nickel-coated low-carbon steel mesh. The anode is made of alkali and oxidation-resistant materials like nickel-coated low-carbon steel or nickel series metals. Electrode catalysts such as Pt on which reaction occurs more easily are sometimes used. A porous diaphragm works for preventing mixture of product gases and direct contact of electrodes. Water decomposes at the cathode to H_2 and OH^-, where the latter migrates through the electrolyte and a separating diaphragm and discharges at the anode, liberating the O_2. The hydrogen is readily extracted from the water when directed into a separating chamber. Operation temperatures are limited to <150°C to avoid corrosion problems; typically, ~90°C is used. The ideal reversible cell potential needed to split water is 1.229 V at

25°C and 0.1 MPa, which corresponds to a theoretical dissociation energy of Gibbs free energy of formation $\Delta G_o = 237$ kJ/mol or an electricity demand of 3.56 kW h/Nm^3 of H_2. However, the realistic cell voltages are 1.7–2.1 V to account for irreversible processes in the reaction mechanism where over-potential of electrodes and ohmic resistance of cell components contribute to the losses. Additional losses are due to gas expansion at the electrodes and to maintain the operation temperature. The electric energy requirement is in the order of 4–4.5 kWh/Nm^3 of H_2, corresponding to an efficiency of 80% and higher. For electrolysis, the theoretical amount of water required is 0.8 L/Nm^3 of H_2; in practice, 1.0 L/Nm^3 is required.

A PEM electrolyze is a PEM fuel cell operating in reverse mode. In PEM electrolysis, water is fed into channels of the anode side of the cell. The water flows from the plate to the anode through the current collector and reacts to split into protons, oxygen, and electrons. Current collectors are porous conductors that allow electrons to transfer from electrode to outer circuit and allow reactant gas from bipolar plate to electrode. The protons migrate through the PEM to cathode side, where it combines with electron to form hydrogen molecule at the cathode. Oxygen gas remains behind in the water. As this water is recirculated, oxygen accumulates in a separation tank and can then be removed from the system. Hydrogen gas is separately channeled from the cell stack and captured. The PEM also serves as a separator of product gases. Solid perfluorosulfonic acid polymer membranes, such as Nafion of DuPont, the United States, are typically used because of its excellent thermal resistance and oxidation resistance. Electrodes come in contact directly with the PEM to avoid interface electric resistance, since corrosion resistance to strong acidity of the PEM is required for electrodes. Typically, platinum group metal, alloy, and oxide of platinum are used for the porous electrodes. In PEM electrolyzer, cathode overpotential is the main source of the total cell overpotential, and it is influenced by selection of material. Typically, oxides of Ir and Ru or metallic platinum are used for cathode material. These materials are often mixed with inert components for structural stability.

The electrolysis is attractive due to its design, simplicity, and the flexibility of accepting virtually any nuclear reactors that generate electricity. The nuclear reactor couples to the process simply via electric transmission, meaning that there is no fluid-thermodynamic connection, such that distributed production of hydrogen close to end user of hydrogen is possible with electricity grids.

4.2.2 Steam Electrolysis (High-Temperature Electrolysis)

High-temperature steam electrolysis (HTSE) operates typically at temperatures 750–950°C and uses steam as feed material. The total energy demand of electrolysis in the vapor phase for this electrolysis is reduced by the heat of vaporization, which can be provided much more inexpensively by thermal rather than electric energy. Since the required electricity input defined by the Gibbs free energy of formation decreases with increasing temperature, HTSE requires about 35% lower electricity compared with conventional electrolysis at low temperatures. Moreover, the efficiency of electricity generation at this high-temperature level is significantly higher (~44%) than standard Rankine-based efficiency of 33%. Thus, HTSE process has overall hydrogen production efficiencies in excess of 50%, compared with 40% with the water electrolysis that is predicted with reactor outlet temperatures above 850°C, when coupled with high-efficiency power cycles such as Brayton cycle. When compared with low-temperature electrolysis, a major difference is that a high-temperature electrolyzer has been coupled with a heat and power source. Nuclear plants, in particular those of the fourth generation, could provide the electricity and, in the case of high-temperature reactors, also deliver relatively high temperatures and high net power cycle efficiencies.

HTSE corresponds to the reverse process of the solid oxide fuel cell. When operated for electrolysis, the cell is referred as solid oxide electrolysis cells (SOEC). Essentially, the electrolytic cell consists of a solid oxide electrolyte with conducting electrodes deposited on either side as shown in schematic Fig. 4.4. The electrolyte is an oxygen-conducting ceramic material, typically yttria (Y_2O_3)-stabilized zirconia (ZrO_2) (YSZ) and MgO. A mixture of steam and small amount of hydrogen is supplied to the hydrogen electrode (cathode) at 750–950°C. HTSE is a combination of two half reactions as shown below:

$$\text{Anode}: 2O^{-} \leftrightarrow O_2\,(g) + 4e^{-}$$
$$\text{Cathode}: 2H_2O\,(g) + 4e^{-} \leftrightarrow 2H_2\,(g) + O_2$$
$$\text{Overall reaction}: 2H_2O(g) \rightarrow 2H_2(g) + O_2(g)$$

On the cathode side, the steam is split into hydrogen and oxygen at the hydrogen electrode-electrolyte interface. On the anode side, oxygen ions are drawn through the ceramic electrolyte by the electric potential of about 1.3 V, where they recombine to oxygen gas at the electrolyte-oxygen electrode interface. The oxygen then flows along the anode, typically made of a

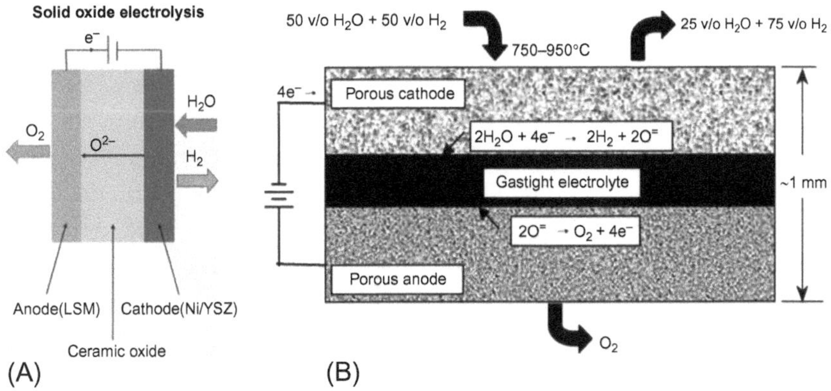

Fig. 4.4 (A) Schematic of solid oxide water electrolysis with an O_2^- conducting electrolyte, a nickel-yttria-stabilized zirconia cathode and a lanthanum strontium manganite anode, and (B) details of the cell and reactions.

composite of YSZ and Sr-doped lanthanum manganite. The hydrogen-steam mixture passes along the hydrogen electrode (Ni/YSZ cermet with ~30% porosity) on the opposite side of the electrolyte. Often, preheated air or steam is used as a sweep gas to remove oxygen from the stack where the sweep gas dilutes the oxygen concentration and thus decreases corrosion of the oxygen-handling components. Pure oxygen can be produced by the stack if corrosion resistant coatings are used for oxygen-handling components. The steam-hydrogen mixture exits from the stack and then passes through a separator to separate hydrogen from the residual steam. The purpose of feed gas stream containing fraction of hydrogen (10%) is to maintain reducing conditions and avoid oxidation of the nickel in the hydrogen electrode.

Since the HTSE cells can be operated at high current densities, large production capacities are possible in comparatively small volumes. A practical electricity-to-hydrogen efficiency of about 90% is achievable with HTSE. Due to the high-efficiency electricity, requirements are 2.6–3 kWh/Nm^3 of H_2 in HTSE. HTSE at about 800°C will have an overall efficiency (including the electric conversion efficiency) in the range of 35%–45%, and the efficiency rises to around 50% at 900°C. HTSE is not yet a commercial technology though there are several efforts in that direction.

Two types of SOEC are available, tubular and planer type. Several experimental studies have been carried out using practical tubular electrolysis cells. In the high operating temperature electrolysis (HOTELLY) program [1], using the 10-cell electrolysis tube, hydrogen was produced at the maximum rate of 6.78 NL/h at 997°C of electrolysis temperature, 370 mA/cm^2 of current density, and a power of 21.7 W. In a stack of HTSE containing 10 modules, each with interconnected 100 cell (total of 1000 electrolysis cells) hydrogen was produced at the maximum rate of 0.6 Nm^3/h [2]. The US Idaho National Laboratory (INL) operated an integrated laboratory-scale (ILS) HTSE unit for 45 days, in 2008 with the use of planar design of SOEC. This demonstration achieved a peak output of 5650 L/h hydrogen. Despite the important milestone, hydrogen production by means of HTE faces challenges, particularly to sustain high-performance operation of electrolyzers. The main issue that needs further improvement remains the lifetime of the hydrogen electrode, which is limited by degradation. Two major problems to be solved are the improvement of the long-duration performance of the cells and the development of techniques for manufacturing and assembling large area cells in order to reduce the overall cost of the commercial plant.

4.2.3 Steam Reforming

The steam reforming process is the catalytic decomposition of light hydrocarbons to react with superheated steam resulting in a hydrogen-rich gas mixture. The main product of steam reforming is synthesis gas, the mixture of hydrogen and carbon monoxide, mainly resulting from the processes of steam reforming, partial oxidation (POX), or autothermal reforming (ATR). The hydrogen fraction is further maximized by water-gas shift reaction following the conversion processes. The final processing step is the separation and purification of the hydrogen. In Fig. 4.5, the steam-methane reforming (SMR) process is shown. A reformer unit includes a desulfurization unit to remove any sulfur impurity in the feed natural gas, the reformer reactor chamber where cleaned natural gas and steam mixture are fed and the catalytic decomposition occurs, the heat exchanger where the product is cooled, a shift-conversion reactor where water-shift reaction takes place, and a hydrogen separator.

The reforming reactions are highly endothermic and take place at high temperatures >500°C. Typically, SMR takes place at 850°C and at pressures >2.5–5 MPa. The energy requirement for the following catalytic decomposition is shown below:

$$CH_4 + H_2O \leftrightarrow 3H_2 + CO \quad -206\,kJ/mol$$
$$CH_4 + 2H_2O \leftrightarrow 4H_2 + CO_2 \quad -165\,kJ/mol$$

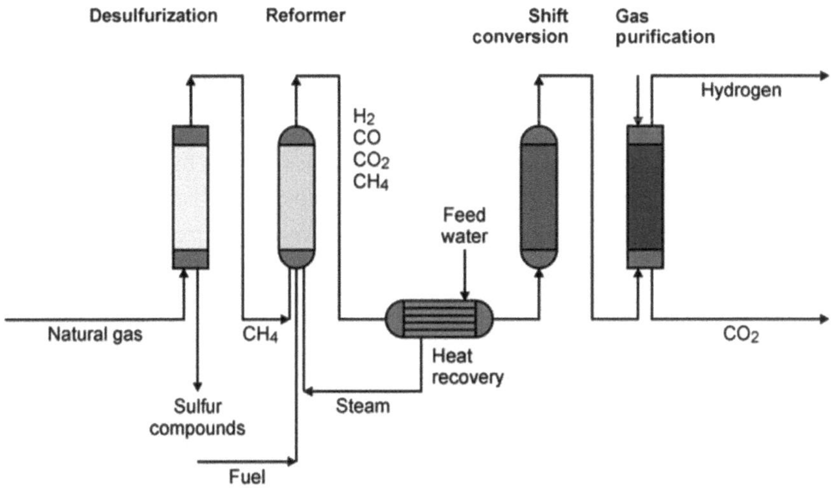

Fig. 4.5 Steam-methane reforming process schematic.

The catalyst used is an iron, nickel, platinum, or ruthenium, which is loaded to an alumina-based material at 10–15 wt%. The feed gas is cleaned of sulfur to protect the catalysts in the reformer. The heat to the reformer tubes in the furnace is provided from outside by burning a part of the natural gas, where heat is transferred by convection and radiation from the flue gas at temperature above 1300°C.

Further, the product CO is converted into CO_2 releasing more hydrogen from steam with a catalytic but slightly exothermic water-gas shift reaction:

$$CO + H_2O \leftrightarrow H_2 + CO_2 + 41\,\text{kJ/mol}$$

Catalysts for this reaction consist mainly of noble metals. This reaction results in more H_2 and a lower CO concentration down to 0.5%–2% of the dry gas and avoids carbon deposition due to the Boudouard reaction. The dry reformer gas without steam on average contains 75% H_2 (a), 13% CO, and 10% CO_2, and remainder 2% is unreformed CH_4. The higher percentages of H_2 and CO are desirable in the reformer gas. This gas is generally referred as synthetic gas that can be tailored by choosing the steam-to-carbon ratio, outlet temperature, and pressure of the reformer, where high steam-to-methane ratio, elevated temperature, and low pressure enable better methane reformation. Synthetic gas or sync gas is used as an intermediate product for the generation of substitute natural gas (SNG), ammonia, or methanol. High steam ratio more than 300% from the stoichiometric mixture favors water-shift reaction with increase in the H_2 yield at 300–400°C. The composition of the synthesis gas is shifted toward a larger CO fraction when the steam is completely or partially replaced by CO_2. The carbon dioxide and other gases are removed from the gas mixture by pressure swing adsorption (PSA) and usually vented to the atmosphere. Tail gases are used for heat requirements such as to elevate feed gas temperature. The hydrogen gas undergoes further purification steps to achieve a purity of >99% before being used, for example, in polymer electrolyte fuel cells. In industrial processes, the shift reaction is generally done through two steps at different temperature levels: the first at 350–450°C with iron-oxide catalyst and the second at 150–250°C with copper-zinc or aluminum catalyst. This results in higher hydrogen concentration and a lower CO concentration, which can be reduced to 0.5%–2% of the dry gas. The CO_2 generated is contained in the PSA reject gas.

Steam reforming of heavier hydrocarbons is possible, but it requires more complex process equipment. Steam reforming of natural gas is a well-established technology on the industrial scale and is the most economic method. It is the current method used in the petrochemical and fertilizer industries to produce hydrogen.

4.2.4 Thermochemical Decomposition of Water

There are number of thermally driven chemical cycles for the decomposition of water to produce hydrogen. These thermochemical cycles include ultra-high-temperature direct thermal splitting of water, high-temperature two-step metal oxide oxidation-reduction reactions to split water, and moderate temperature three-step water-splitting cycles. The most direct and simplest thermochemical process to decompose water is the one-step thermolysis of the water molecule at elevated temperatures and separating the hydrogen from the equilibrium mixture. However, the decomposition of water does not proceed well until the temperature is very high. The Gibbs free energy change ΔG does not become zero until the temperature is increased to about 4700 K. The issues with materials capable of withstanding such high temperatures required for direct water thermolysis make this process economically prohibitive. At 25°C and 1 atm condition (STP), the property changes for water decomposition $H_2O(l) \rightarrow H_2(g) + ½O_2(g)$ are given below:

$$\text{Enthalpy change } \Delta H = 68.3 \text{kcal/gmol}$$
$$\text{Free energy change } \Delta G = 56.7 \text{kcal/gmol}$$
$$\text{Entropy change } \Delta S = 0.039 \text{kcal/gmolK}$$

The ideal work requirement in an electrochemical process, such as water electrolysis, is equal to the change in the free energy [3]. The entropy change is the negative of the temperature derivative of the free energy change, and it is too small to make direct decomposition feasible. The ideal work requirement in a chemical process comprising several chemical reactions is the sum of the positive free energy changes. Thus, reducing the free energy change is the same as improving the reaction equilibrium. The work requirement for a multireaction process could be reduced to zero at reasonable operating temperatures with some constrains on the chemical reactions. Running certain reactions at elevated temperatures, where ΔG is zero and the others are at lower temperatures, enables efficient cycle for decomposition in multi-reaction process. Since the 1960s, hundreds of efficient multistep cycles for decomposition of water have been examined where the efficiency measure was usually taken to be the higher value of the hydrogen made, divided by the heat required to make it. International efforts have led to identification of numerous thermochemical cycle-based criteria such as thermal efficiency, conversion of chemical reactions, side reactions, toxicity of elements and compounds involved, maximum process temperature, material separation, materials for handling, and cost; some cycles have been found to be most practical.

Some cycles have been systematically studied for their scientific and practical feasibility and have been experimentally demonstrated to certain extent. Among well-studied thermochemical cycles, the sulfur family cycles have been identified as the potentially promising candidates with higher efficiency and a lower degree of complexity in terms of number of reactions and separations. Fig. 4.6 shows these three sulfur cycles: sulfur-iodine cycle, hybrid sulfur cycle, and sulfur-bromide hybrid cycle with their reaction temperatures. The sulfur-iodine (SI) or iodine-sulfur (IS) cycle is currently the leading candidate, which has been studied extensively. Fig. 4.7 presents the simple schematic of the SI cycle. The central Bunsen reaction produces two kinds of acid, sulfuric acid (H_2SO_4) and hydriodic acid, in solution (HI) from sulfur dioxide (SO_2), iodine (I_2), and water (H_2O). The mixed acid separates into two types of acid of its own accord (liquid-liquid separation). The acid, which is rich in HI, is HIx phase (HIx solution), while the acid,

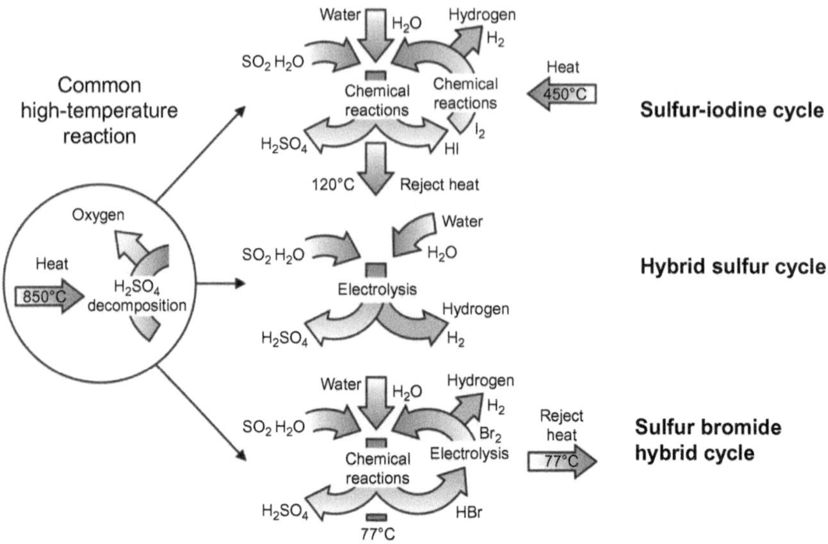

Fig. 4.6 Sulfur family thermochemical cycles.

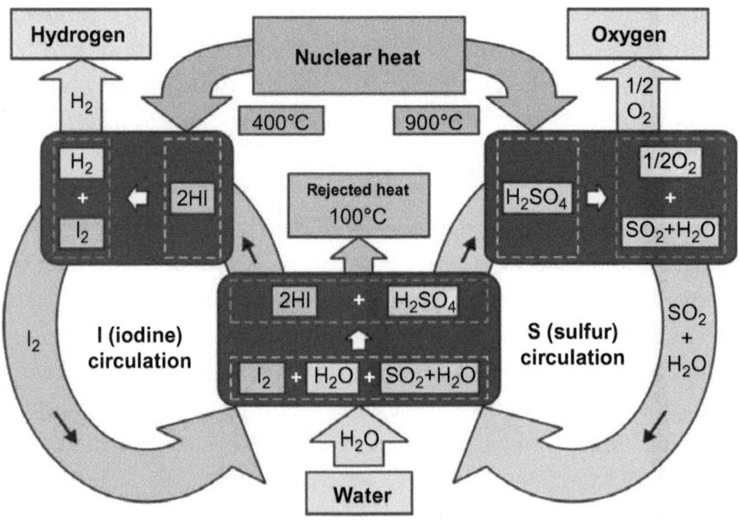

Fig. 4.7 Scheme of the sulfur-iodine cycle.

which is rich in H_2SO_4, is the H_2SO_4 phase. After the separation of the acids, they are purified, concentrated, and decomposed in the other two reactions. The H_2SO_4 decomposition reaction produces oxygen, sulfur dioxide, and water. The HI decomposition reaction produces hydrogen and iodine. With the exception of hydrogen and oxygen, the other products can be reused in the Bunsen reaction step as the reactant material. The endothermic H_2SO_4 decomposition reaction can be operated at about 800–1000°C. The decomposition of hydriodic acid involves an endothermic reaction around 400–500°C. The Bunsen reaction occurs exothermically at temperatures of about 100°C. The SI cycle has been studied extensively with small-scale test facilities for the demonstration of the technology in the United States, Japan, Europe, South Korea, and China.

The hybrid sulfur (HyS) process also known as Westinghouse cycle is a variation of the SI cycle process with only two reaction steps, sulfuric acid decomposition and electrodialysis of SO_2:

$$(H_2SO_4)g \rightarrow (SO_2)g + (H_2O)l + 0.5O_2 \quad (\sim 850^{\circ}C)$$
$$(SO_2)g + (2H_2O)l \rightarrow H_2SO_4 + H_2 \text{ electricity} \quad (20 - 110^{\circ}C, 0.17V, \text{ and } 0.2 - 1\,MPa)$$

The sulfur-bromine hybrid cycle uses bromine instead of iodine. The reactions steps are as follows:

$$H_2SO_4 \rightarrow H_2O + SO_2 + 0.5O_2 \quad (700 - 1000^{\circ}C)$$
$$SO_2 + Br_2 + 2H_2O \rightarrow 2HBr + H_2SO_4 \quad (20 - 100^{\circ}C)$$
$$HBr \rightarrow H_2 + Br_2 - \text{electricity} \quad (80 - 200^{\circ}C)$$

Other important thermochemical and hybrid cycles are listed in Table 4.2. The temperatures and if electricity was used for each reaction are indicated in the table. More information on these and other cycles are available from references [3].

Thermochemical processes have proved to be challenging. Though many challenges have been resolved through research and development work, many still remain to be resolved. Many chemical reactions have low equilibrium conversion ratio because of their positive Gibbs energy, such as HI decomposition in IS process and $FeCl_3$ decomposition in Fe-Cl processes that generates great amount of recycling flow and low thermal efficiency. Many thermochemical processes include severe environment for component materials. Materials with high resistance to corrosion are required. Separation of objective product, by-products, and nonreacted reactants has some difficulties, especially in the mixture of the same phase. When separation of objective product is not enough, recycling of the flow is required, and flow rate becomes larger. Thermal efficiency improvement process is required. Today, thermal efficiencies based on practical assumptions reach around 35%–45%.

Table 4.2 Other Thermochemical and Hybrid Cycle Studied in Literature

Calcium-bromine (UT-3) cycle	
$(CaBr_2)s + (H_2O)g \rightarrow (CaO)s + (2HBr)g$	700–750°C
$(CaO)s + (Br_2)g \rightarrow (CaBr_2)s + (0.5O_2)g$	500–600°C
$(Fe_3O_4)s + (8HBr)g \rightarrow (3FeBr_2)s + 4H_2O + (Br_2)g$	200–300°C
$(3FeBr_2)s + 4(H_2O)g \rightarrow (Fe_3O_4)s + (6HBr)g + (H_2)g$	550–600°C
Hybrid Ca-Br cycle	
$CaBr_2(s) + H_2O(g) \rightarrow CaO(s) + 2HBr(g)$	750°C
$CaO(s) + Br_2(g) \rightarrow CaBr_2(s) + 0.5O_2(g)$	580°C
$2HBr(g) \rightarrow H_2(g) + Br_2(g)$ – electricity	100°C
Iron-chlorine cycle	
$3FeCl_2(s) + 3H_2(g) \rightarrow 6HCl(g) + 3Fe(s)$	877°C
$3Fe(s) + 4H_2O(g) \rightarrow Fe_3O_4(s)\ 4H_2(s)$	327°C
$H_2O(g) + Cl_2(g) \rightarrow 2HCl(g) + 0.5O_2(g)$	1027°C
$Fe_3O_4(s) + 8HCl(g) + 0.5Cl_2(g) \rightarrow 1.5Fe_2Cl_6(s) + 4H_2O(g)$	427°C
$1.5Fe_2Cl_6(s) \rightarrow 3FeCl_2(g) + 1.5Cl_2(g)$	327°C
Copper-chlorine cycle (five step)	
$(2Cu)s + (2HCl)g \rightarrow (2CuCl)l + (H_2)g$	430–475°C
$(4CuCl)aq \rightarrow (2CuCl_2)aq + (2Cu)s$ electricity	30–70°C
$(2CuCl_2)aq \rightarrow (2CuCl_2)s$ drying	<100°C
$(2CuCl_2)s + (H_2O)g \rightarrow (CuO{\cdot}CuCl_2)s + (2HCl)g$	400°C
$(CuO{\cdot}CuCl_2)s \rightarrow (2CuCl)l + (0.5O_2)g$	500°C
Copper-chlorine cycle (four step)	
$2CuCl(aq) + 2HCl(aq) \rightarrow 2CuCl_2(aq) + H_2(g)$ electricity	70°C
$2CuCl_2(aq) \rightarrow 2CuCl_2(s)$	70°C T
$2CuCl_2(s) + H_2O(g) \rightarrow CuO{\cdot}CuCl_2(s) + 2HCl(g)$	400°C
$CuO{\cdot}CuCl_2(s) \rightarrow 2CuCl(l) + 0.5O_2(g)$	500°C

4.2.5 Carbon, Hydrocarbon and Biomass Conversion

Fossil fuel (coal and hydrocarbons) and biomass can be converted into H_2 and CO via steam reforming or gasification. The terms steam reforming and gasification are referred to as conversion with steam of vaporous fuel such as methanol into syngas and that of solid fuel such as coal, respectively. Coal is one of the abundant fuels and the major source of energy, accounting for about 25% of the world's primary energy consumption and 40% of the world's electricity production. Gasification of coal being the oldest hydrogen production technology today, the conversion of coal to gaseous or liquid fuels is being applied commercially worldwide. Coal gasification is a clean coal technology producing synthesis gas and hydrogen and with the potential for easy separation of CO_2 stream. Coal is a solid with a high carbon content and an approximate 5% content of hydrogen $\sim(CH_{0.8})_n$.

The conversion of coal into a gas is realized by means of a gasification agent, steam, oxygen, air, hydrogen, carbon dioxide, or a mixture of these, which reacts with the coal at temperatures >800°C, similar to an incomplete combustion. If air or oxygen is injected into the gasifier, a part of the coal is directly burned (POX), allowing for an autothermal reaction. In the coal gasification with steam, two consecutive processes take place, the POX or pyrolysis reaction and the water-gas reaction:

$$C + H_2O \leftrightarrow H_2 + CO \quad -118.5\,kJ/mol\,C \quad 400-600^{\circ}C$$
$$CO + H_2O \leftrightarrow H_2 + CO_2 \quad +43.3\,kJ/mol\,C$$

In the pyrolysis reaction, all volatile constituents of the coal are rapidly expelled. The gasification reaction with the steam is through heterogeneous water-gas reaction and the homogeneous shift reaction with a further increase of the H_2 fraction. In this gasification reaction, residual organic solids are converted to synthesis gas with some CO_2 and steam. If sync gas is the end product, then this is followed by a methanation step typically carried at high temperatures and low-pressure reactions:

$$CO + 3H_2 \leftrightarrow CH_4 + H_2O \quad +206.0\,kJ/mol\,C$$

Biomass is generally classified into massive biomass (dedicated bioenergy crops, process residues, harvest residues, and energy crops) and organic biomass (liquid manure, sludge, and process residues). The massive biomass can be directly converted to hydrogen with thermochemical process. However, organic biomass is subject to anaerobic fermentation to produce methane-rich gas and then into hydrogen by reformation. The amount of hydrogen in biomass is in the range of 6–7 wt%. Plant-originated biomass resources, woody biomass, and herbaceous one consist of the three major chemical constituents, cellulose, hemicellulose, and lignin. Steam gasification of biomass is generally conducted in two steps. First, thermochemical decomposition of the biomass takes place with cracking and reforming of volatiles and the production of tar and char and gasification of the char. In the second step, hydrocarbon gases and carbon in the biomass react with CO, CO_2, H_2, and H_2O, leading to more light gases. The pyrolysis is the primary step of biomass gasification that involves degradation, the polymers evolving their fragments generally called volatiles, and repolymerization/carbonization forming carbonized solid termed char. Volatiles consist of inorganic gases, light hydrocarbon gases, and condensable organic compounds that are called tar. The volatiles released into the vapor phase undergo thermal cracking (secondary pyrolysis), steam reforming, and/or O_2 (partial or full combustion). The thermal cracking produces not only light gases such as H_2, CH_4, and CO but also refractory monoaromatic compounds such benzene and phenol and polyaromatic hydrocarbons in parallel. Dry gas yields carbon conversion efficiency, and the hydrogen fraction in the product gas increases with increasing temperatures from ~25% to ~50% if the temperature increases from 600°C to 900°C, respectively. The char experiences partial/full combustion or steam gasification reacting with O_2 and steam, respectively. The steam gasification of char is in general the slowest reaction among those involving the pyrolysis products, and hence, it often determines overall rate of the biomass conversion into gases.

The organic substances are decomposed with gasification or pyrolysis and subsequent autothermal or allothermal steam reforming. The gasifiers are usually indirectly heated or oxygen blown to avoid nitrogen in the product gas and are operated at low pressures. The autothermal gasification in a fluidized bed results in a synthesis gas with typically 30% H_2, 30% CO, 30% CO_2, and 5%–10% CH_4 plus some higher hydrocarbons. A shift reaction again converts CO to increase the hydrogen fraction. Anaerobic fermentation of wet biomass leads to a CH_4-rich gas with only little hydrogen.

Reformation of common hydrocarbons methanol, ethanol, dimethyl ether, and benzene to produce hydrogen is listed in Table 4.3. Production of hydrogen from methanol is usually done either by POX or steam reforming of methanol. POX is characterized by an exothermic reaction and high reaction rates. Steam reforming of methanol is endothermic and takes place at moderate temperatures in the range of 250–350°C. Ethanol being renewable has advantage as being CO_2 neutral. Catalytic reforming of ethanol provides total of 6 mol of H_2 per mole of ethanol in the endothermic process. Dimethyl ether, generally produced from methanol, is an organic compound with properties similar to liquid petroleum gas. The steam reforming of dimethyl ether produces hydrogen at temperatures of 285–300°C.

Table 4.3 Reformation of Hydrocarbons for Hydrogen Generation

Methanol	$CH_3OH + H_2O \leftrightarrow CO_2 + 3H_2$	250–350°C
Ethanol	$CH_3CH_2OH + 3H_2O \leftrightarrow 2CO_2 + 6H_2$	−173.4 kJ/mol
Dimethyl ether	$CH_3OCH_3 + H_2O \rightarrow 2CH_3OH$	
	$2CH_3OH + 2H_2O \rightarrow 6H_2 + 2CO_2$	
Benzene	$C_6H_6 + 12H_2O \leftrightarrow 6CO_2 + 15H_2$	−458 kJ/mol

4.2.6 Radiolysis of Water

The interaction of various types of ionizing radiation (α, β, and γ) with water produces molecular hydrogen. The radiolytic method is used for several industrial applications, for example, for sterilization, food irradiation, radiotherapy, polymer production, and water remediation. Fig. 4.8 shows schematic of the radiolysis of water molecule. The splitting of the water molecule occurs through excitation or ionization of water molecule. The ionization leads to H_2O^+ cation and an electron e^-, where the cation reacts with a surrounding H_2O to form a hydroxyl radical, OH, and the ejected e^- forms a hydrated electron with surrounding H_2O. The partially excited H_2O decomposes into H and OH (or $H_2 + O$). The products of radiolysis of water are e_{aq}^-, H, OH, H_3O^+, H_2, and H_2O_2. As the OH radical is an oxidizing agent, it abstracts electron or hydrogen from many ions and molecules. In contrast, the e_{aq}^- and H atom are reducing agents. The e_{aq}^- reacts by electron attachment with many kinds of solutes, and the H atom can also abstract hydrogen especially from saturated organic molecules to give H_2. After the sequence of many reactions of radiolysis products, radical species are consumed to give molecular products. H_2 and H_2O_2 are the main products in radiolysis of pure water.

The generated hydrogen from the irradiation of water with β and γ radiation is low, <1 mol per 100 eV of absorbed energy, and this is largely due to the rapid reassociation of the species arising during the initial radiolysis. When the radiation deposits energy if impurities are present or if physical conditions are created that prevent the establishment of a chemical equilibrium, the net production of hydrogen can be greatly enhanced. The yield of

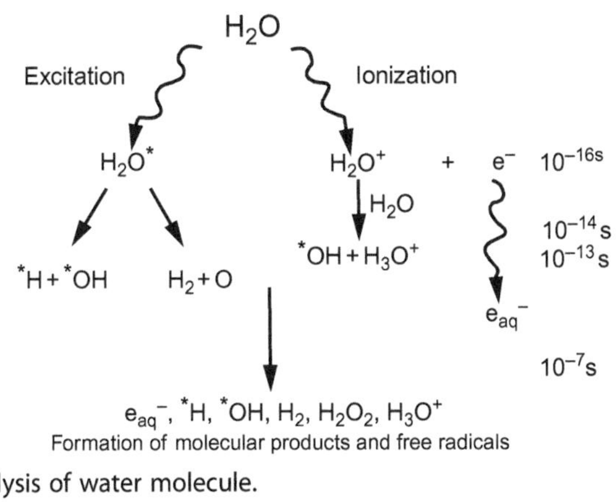

Fig. 4.8 Radiolysis of water molecule.

H_2 was found to be greater for the radiolysis of water adsorbed on different solid surfaces than in aqueous solution and bulk water.

4.3 NUCLEAR SYSTEMS FOR HYDROGEN PRODUCTION

Currently, 450 commercial reactors are operating mainly producing electricity. Table 4.4 lists the type of commercial operating reactors and their numbers, type of fuel used, and the amount for power generation. Among these thermal reactors cooled by light water or heavy water referred as generation three, Gen III reactors form the majority. Most of Gen III reactors operate with power rating from 500 to 1000 MWe. The light-water and heavy-water reactors operate in the temperature range of 270–330°C. The commercial reactors generate electric power using the nuclear heat energy through Rankine cycle. The hydrogen can be produced with nuclear reactor heat or heat and electricity as described in hydrogen production methods through electrolysis, steam reforming, or thermochemical or hybrid processes. The overall process efficiency of nuclear heat to hydrogen

Table 4.4 Commercial Operating Nuclear Power Plans (Data as of November 2016)

Operational Reactor Type		Number	Fuel	Electricity Net Output (GWe)
PWR	Pressurized light-water-moderated and light-water-cooled reactor	291	Enriched UO_2	272.765
BWR	Boiling light-water-cooled and light-water-moderated reactor	78	Enriched UO_2	75.208
PHWR	Pressurized heavy-water-moderated and heavy-water-cooled reactor	49	Natural UO_2	24.634
GCR	CO_2 gas-cooled, graphite-moderated reactor	14	Natural UO_2 (metal) Enriched UO_2	7.720
LWGR	Light-water-cooled, graphite-moderated reactor	15	Enriched UO_2	10.219
FBR	Fast breeder reactor, liquid sodium-cooled reactor	3	PuO_2 and UO_2	1.369
Total		450		391.915

Source: IAEA, Nuclear power reactors in the world, IAEA Report, IAEA-RDS-2/37, ISBN 978-92-0-104017-6 ISSN 1011-2642, May 2017.

production is about 25% with today's light-water reactors with 33% for reactor electric power generation for water electrolysis. Several new-generation reactors called Gen IV operate at high temperature up to 1000°C. These are listed in Table 4.5. These high-temperature reactors can produce hydrogen up to 50% efficiency using the combination of steam electrolysis or a thermochemical or hybrid process. All proposed Gene IV systems are able to supply heat for lower-temperature industrial processes ($<\sim$500°C).

4.3.1 Light and Heavy Water Reactors

There are over 450 commercial light-water reactors (LWRs) in operation and further 60 under construction to date. Light water (H_2O) is used as both coolant and neutron moderator in the reactor that operate with low-energy range (thermal) neutrons. The controlled fission heat that is released in the core is removed by the coolant water from the core to the balance of plant for energy conversion and delivery. The reactor core consists of nuclear fuel assembles made of fuel rods encased in metallic cladding, reactivity control rods, and support structure. Both ends of the cladding tube containing high-pressure helium gas are welded to seal the gas in the tube. The fuel rods contain many small pallets and are made of low-enriched (<5%) sintered uranium dioxide. A fuel assembly may contain as many as 200 fuel rods arranged in square array, and hundreds of assemblies may be loaded in a large-size reactor core. As the fuel undergoes fission (burned), it is replaced with fresh fuel normally in <2 years. A steel pressure vessel contains the reactor core and is designed to withstand all thermal and seismic loadings and irradiation exposure over the life span of the reactor operation in the range of 40 years. The control rods are made of materials that absorb neutrons such as boron carbide (B_4C), hafnium (Hf), or their combinations. The core has a number of control rods distributed in each fuel assembly to control neutron flux and hence fission process and reactor power. Inserting the control rods deeper into the core causes more neutrons absorbed and consequently stopping the chain reaction. Pulling the control rods away from the core results in more neutrons absorbed in the core and hence accelerates the fission. There are two types of commercial LWR, pressurized water reactor (PWR) and boiling water reactor (BWR). A PWR has two coupled main cooling systems, primary and secondary as shown in Fig. 4.9.

The primary systems consist of the pressure vessel and the coolant flow path from reactor vessel where water is heated and flow through hot leg to steam generator (secondary cooling system) where the heat is transferred and return to the pressure vessel through the cold leg. A pressurizer helps to

Table 4.5 Gen IV Reactor Designs Under Development by GIF

	Neutron Spectrum (Fast/ Thermal)	Coolant	Temperature (°C)	Pressure[a]	Fuel	Fuel Cycle	Size (MWe)	Use
Gas-cooled fast reactors (GFR)	Fast	Helium	850	High	U-238[b]	Closed, on site	1200	Electricity and hydrogen
Lead-cooled fast reactors (LFR)	Fast	Lead or Pb-Bi	480–570	Low	U-238[b]	Closed, regional	20–180[c] 300–1200 600–1000	Electricity and hydrogen
Molten-salt fast reactors (MSR)	Fast	Fluoride salts	700–800	Low	UF in salt	Closed	1000	Electricity and hydrogen
Molten-salt reactor—advanced high-temperature reactors (AHTR)	Thermal	Fluoride salts	750–1000		UO_2 particles in prism	Open	1000–1500	Hydrogen
Sodium-cooled fast reactors (SFR)	Fast	Sodium	500–550	Low	U-238 and MOX	Closed	50–150 600–1500	Electricity
Supercritical water-cooled reactors (SCWR)	Thermal or fast	Water	510–625	Very high	UO_2	Open (thermal) closed (fast)	300–700 1000–1500	Electricity
Very-high-temperature gas reactors (VHTR)	Thermal	Helium	900–1000	High	UO_2 prism or pebbles	Open	250–300	Hydrogen and electricity

[a]High = 7–15 MPa.
[b]With some U-235 or Pu-239.
[c]"Battery" model with long cassette core life (15–20 years) or replaceable reactor module.
Table adapted from http://www.usnuclearenergy.org/GENIVReactors.htm.

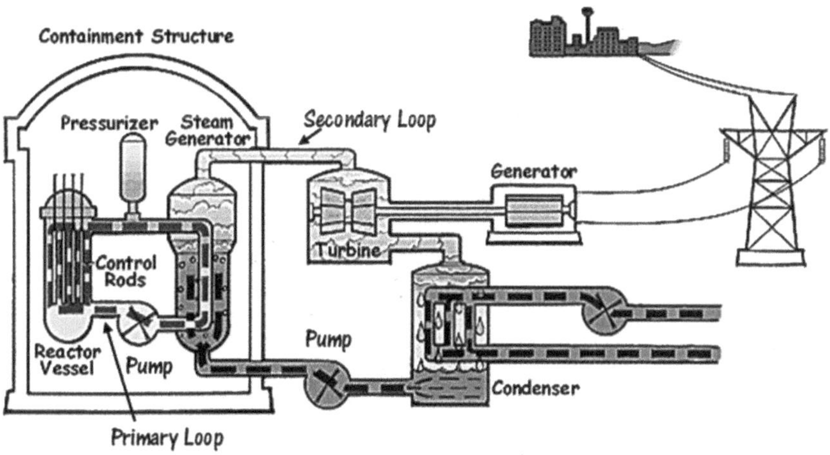

Fig. 4.9 The pressurized water reactor.

maintain compressible volume in the primary cooling system, and a pump in the cold leg circulates the water. The water in the primary system is maintained at high pressure (15 MPa) and in subcooled condition. The secondary cooling system consists of the steam generator, main steam lines, generators, condenser, pump, and feedwater lines. The steam flows through main steam lines to turbine where power is generated, and the spent steam is condensed, and water is returned to steam generator through feedwater line. The primary water coolant exits the reactor core at a temperature of about 330°C before flowing to the steam generator under the circulation of the primary coolant pump. In the steam generator, the primary coolant transfers heat to generate steam of about 287°C and 7 MPa in the secondary coolant system. The steam expands in the turbine for electric power conversion, is condensed to a liquid in the condenser, and pumped back to the steam generator. Typically, seawater or lake water is used to remove the thermal waste heat of the secondary system power conversion in the condenser. The primary system and the steam generators are housed in a large building called containment that isolates these systems from environment in case of an accident.

In contrast to PWR, the BWR has just one primary cooling system. The water is heated, boiled, and converted to steam in the reactor core itself. The reactor consists of a steel reactor pressure vessel, recirculation pumps, and a pressure suppression pool inside a reactor containment vessel and the turbine power generator system installed outside of the containment vessel as shown in Fig. 4.10. The steam generated in the core is sent through main steam lines

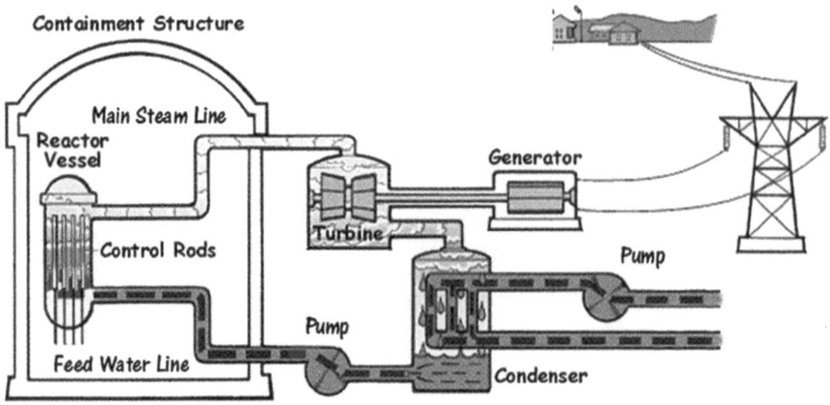

Fig. 4.10 The boiling water reactor.

to drive the turbine power generator. The turbine exhaust steam is condensed by external cooling of typically seawater or lake water. The water is returned to the reactor by the feedwater pump. The steel reactor vessel structurally supports and encloses the reactor core, steam separator, and dryer located top of the core. The reactor core contains mainly of fuel assemblies and control rods. The design of the fuel assembly and control rods in BWR are slightly different from PWR.

A fuel assembly for BWR consists usually of a square array of fuel rods. Similar to PWR, BWR fuel rod consists of a zircaloy cladding tube containing UO_2 ceramic pellets. The fuel assembly in BWR is in closed canisters in contrast to PWR fuel assemblies that are open. The control rods are cruciform-shaped blades located between fuel assemblies. The coolant water is boiled to saturation at 7 MPa to generate steam at 286°C. About 40% of water is boiled into steam, and remaining water is recirculated with recirculation pumps. Like in the PWR, control rods are inserted into or pulled away from the core to control the core reactivity and power generation rate. In addition, the BWR power generation is also controlled by adjusting the coolant-recirculation flow through the reactor core by the recirculation pumps.

Recently, advanced designs of PWR and BWR are being commercially advanced, and some of them are being constructed. These referred as Gen III+ systems are ABWR and ESBWR in BWR category and AP1000, US-APWR, and EPR (European pressurized reactor) in PWR categories. The advanced reactors have typically large power ratings up to 1600–1700 MWe and have passive safety systems for enhanced safety during accident scenarios.

The heavy-water (D_2O)-moderated and heavy-water-cooled thermal neutron reactor (HWR) design referred as CANDU reactor is the world's third most common type of commercial reactor. CANDU has different core design where instead of single pressure vessel, fuel bundles are stacked inside pressure tubes that are placed horizontally parallel in a vessel called calandria (Fig. 4.11). This design enables online refueling in the individual pressure tubes continuously throughout reactor operation. The calandria encases the coolant channels containing fuel bundle, moderator, and control systems. Separate flow loops of heavy water (D_2O) are used for coolant at high temperature in the pressure tube and for moderator in low temperature and pressure in the calandria. Due to more efficient neutron moderation by heavy water, these reactors need low-enriched uranium unlike the LWRs and use fuel of natural uranium containing <1% of fissile uranium. The reactor coolant is heated up to 313°C in subcooled condition at 10.6 MPa. This reactor has primary and secondary coolant loops similar to PWR. The primary heat is transferred to secondary in steam generator to produce steam to 263°C and 4.9 MPa that drives steam turbine to generate electricity. Currently, more than 30 of CANDUs are operating in the world, and there are additional 13 CANDU-type reactors in India. Advanced designs including CANDU-6 and Generation III+ ACR-1000, rated at 700–1100 MWe, are available for new construction.

The current LWR and HWR fission-heat energy can be readily used with large electrolyzer units to produce hydrogen. As these power plants serve as base load for electric supply, these reactor energy applications can be extended by using lower-cost off-peak electricity for hydrogen production and to store nuclear power in the form of hydrogen for the shaving of

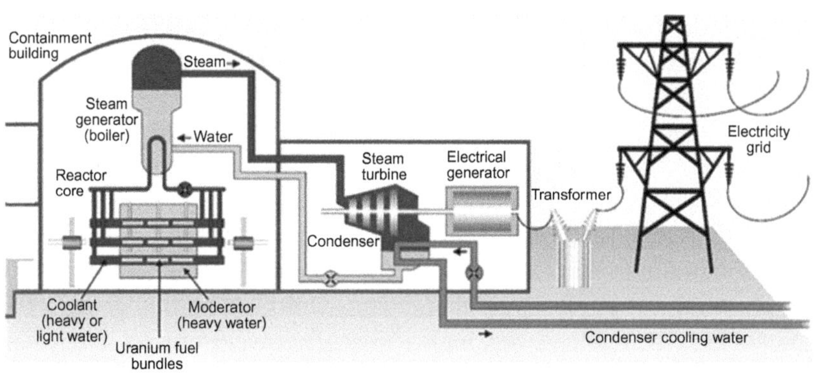

Fig. 4.11 Schematic of CANDU power plant.

peak electricity demands with efficient power conversion technologies such as fuel cells and combustion turbine electric generator. There are several economic incentives for these applications.

4.3.2 High-Temperature Gas Cooled Reactors

As listed in Table 4.4, the gas-cooled reactor (GCR) called advanced gas-cooled reactor (AGR) is CO_2 cooled and graphite-moderated thermal reactor. In the United Kingdom, there are currently seven AGR nuclear power stations (five in England and two in Scotland) each with two operating reactors. This reactor operates at maximum coolant temperature of 336°C. The reactor core, boilers, and gas circulators are housed in a single prestressed concrete cavity known as the pressure vessel. The reactor core is a 16-sided stack of interconnected graphite bricks maintained in position by a steel restraint structure surrounding the core and is supported by a system of steel plates. Heat is produced within the core of the reactor and is transferred to pressurized carbon dioxide gas, which is pumped over the fuel by gas circulators around the gas baffle. The gas, in turn, transfers its heat to boilers. Feedwater is pumped into the boilers and is converted to superheated steam, which is taken to a steam turbine generator.

In contrast to this, the high-temperature gas-cooled reactors (HTGRs) are typical helium-cooled and graphite-moderated reactor operating in thermal neutron spectrum. It is also referred as *very-high-temperature gas reactor* (VHTR), one of Gen IV reactor with the coolant outlet temperature at 900–1000°C. The reactor core type of the VHTR can be a prismatic block core or a pebble-bed core. The prismatic block core consists of hundreds of graphite blocks with fuel rods and coolant channels distributed throughout each block. The design of VHTR with pebble core referred as pebble-bed modular reactor (PBMR) consists of random packing of pebbles of 60 mm diameters made of high-density graphite and fuel particles. The pebbles in the packing move freely downward due to gravity. This feature enables online refueling of PBMR. The fuel rods and pebbles are made of fuel particles coated with strong ceramic (SiC or ZrC) making individual spherical shells of 1 mm diameter with maximum design limit of 1600°C. These particles are referred as TRISO fuel particles. The designs usually select once through uranium fuel cycle; however, multifuel cycle is also being considered due to the strong ceramic fuel design with high burnup ability. The design of HTGR is to have very high degree of passive safety to avoid release of fission products under all conditions of normal operation and accidents, including beyond design basis events.

For electricity generation, the helium gas turbine system (Brayton cycle) can be directly interfaced with the primary coolant loop allowing high efficiency of more than 50%. The high outlet coolant temperature HTGR enables hydrogen production with conventional electrolysis with overall thermal efficiency of 35%–40%. The reactor can be deployed for high-temperature steam reforming of biomass, coal, and natural gas, as has been demonstrated in Japan and Germany. The HTGR system is considered as the prime candidate for large-scale hydrogen production and for cogeneration of heat and power for large industrial complexes. HTGR could be deployed in refineries and petrochemical industries to substitute large amounts of process heat at different temperatures, including hydrogen generation for upgrading heavy and sour crude oil. Core outlet temperatures of 1000°C and higher would expand nuclear heat applications to such processes as steel, copper, and aluminum production. Heat application processes are generally coupled with the reactor through an intermediate heat exchanger. The hydrogen production of HTGR in combination with high-temperature steam electrolysis and thermochemical and hybrid processes is being studied worldwide.

4.3.3 Liquid Metal Cooled Reactors

Liquid metal cooled reactors have fast-neutron spectra as they lack moderator and are classified based on the liquid metal coolant used such as sodium, lead-bismuth eutectic alloy, and lead bismuth. The key feature of fast reactors is their ability to breed fissile material from fertile material such as 238 U and 232Th into Pu239 and U233, respectively, and manage high-level wastes and particularly plutonium and other actinides. Few sodium-cooled fast reactors (SFRs) have been designed and operated in the range of medium size (150–500 MWe) to large size (above 1000 MWe). The SFR core outlet temperatures are typically 530–550°C. The fuel forms are either mixed oxide fuels (U-Pu) or U-Pu-Zr metal alloy alone or metal alone or with minor actinides for waste destruction. The fuel cycle in SFR requires reprocessing of spent fuel. Besides good heat transfer at the core and a large margin to coolant boiling, a major safety feature with SFR is that the primary system operates at essentially atmospheric pressure, pressurized only to the extent needed to move the fluid. Demonstration plants in several countries ranged from 1.1 MW(th) (EBR-I in 1951) to 1200 MW(e) (Superphénix in 1985). Today, SFRs are being operated in India and China.

The lead- or lead-bismuth alloy-cooled fast reactors (LFR), Gen IV reactors, also have closed fuel cycle for efficient conversion of fertile fuel and actinide management. They have low power density and are natural circulation cooled reactors with fissile self-sufficient core designs for very long refueling interval. The plant sizes include small battery type of systems thermally rated at 50 MWe–150 MWt with long refueling interval (10–30 years core life), medium (300–400 MWe) and large (1200 MWe). For the fuel, the nearer term options use metal alloy fuel or nitride fuel, containing fertile U and transuranics (TRU). The technology from sodium-cooled reactor system on metal alloy fuel pin performance at 550°C and U/TRU/Zr metal alloy recycling, and remote fabrication technologies can be easily adapted for these reactors.

The advanced LFR outlet coolant temperatures range from 550 to 800°C, depending on the selection and feasibility of structural materials and fuels. The broad coolant-temperature range can be used for hydrogen production through thermochemical process and hence is advantages than with the sodium reactor.

4.3.4 Gas-Cooled Fast Reactors

These are Gen IV reactors similar to gas-cooled high-temperature reactors under development. The gas-cooled fast reactor (GFR) system was heavily pursued in the 1970s by companies like General Atomics; however, no plants utilizing the GFR concept were ever built. Recently, the designs have been proposed for modular and larger systems rated between 600 and 2400 MWt for high-temperature operation to 850°C. They use helium as coolant, which is neutron transparent as is preferable with fast-neutron spectrum, with reactor outlet temperature of 850°C. Helium is inert and benign with structural materials. Several fuel forms are being considered for their potential to operate at the high temperatures and to ensure excellent retention of fission products: composite ceramic fuel, advanced fuel particles, or ceramic clad elements of actinide compounds and even depleted uranium. The GFR minimizes production of long-lived radioactive waste. This is possible not only because of the fast spectrum but also because of the recycling of all actinides. The used fuel would be reprocessed onsite and all the actinides recycled repeatedly to minimize production of long-lived radioactive waste. Core configurations are being considered based on pin- or plate-based fuel assemblies or prismatic blocks. The major development activities for the GFR are concentrated in France, Japan, and more recently in the United States. In the United States, a helium-cooled fast reactor proposal, called the Energy Multiplier Module (EM2), uses as its fuel the spent fuel from existing reactors without prior reprocessing. Many of the technology requirements such as helium coolant and direct cycle gas turbine balance of plant are shared with another long-running thermal-neutron reactor design, the GT-MHR. EM2 is rated at 500 MWt with coolant outlet temperature at 850°C. The core life span of 30 years is based on spent LWR fuel and depleted uranium (tails of uranium enrichment) without the need for refueling. If realized, EM2 would offer a unique advantage of nuclear fuel recycle without reprocessing. A key design advantage for the GFRs is the high-temperature nuclear heat capability as a result of selecting helium as coolant. At 850°C, it can drive a direct Brayton cycle gas turbine for high-efficiency power production and open the full range of possible routes for hydrogen production as indicated in Table 4.5. In particular, the HTE and thermochemical process can be supported by this reactor for highly efficient hydrogen production.

4.3.5 Molten Salt Reactors

Two types of these salt-cooled reactor proposals are characterized by fuel forms. The earlier proposal known simply as molten-salt reactor (MSR) features fuel-coolant liquid mixture of sodium, zirconium, and uranium fluorides. The homogenous liquid fuel allows various mixing of fuel and actinides for waste destruction. The flexibility with liquid fuel eliminates the need for fuel fabrication. The molten-salt fuel flows through graphite core channels, producing an epithermal to thermal spectrum. The more recent design, called advanced high-temperature reactor (AHTR), uses coated fuel particles with low-enriched uranium kernel, and a molten salt serves as coolant only. The fuel particles are dispersed in graphite matrix. The core is of thermal spectrum. The AHTR's molten-salt coolant is a mixture of fluoride salts with melting point near 400°C and atmospheric boiling point around 1400°C. Various salts have been evaluated such as $7Li_2BeF_4$ and $NaF\text{-}ZrF_4$. The reactor operates at low pressure (<0.5 MPa) with the core outlet coolant temperatures in a range of 700–1000°C. The reactor rating falls in the range of 900–2400 MWt. Heat is transferred from the primary coolant through a compact secondary molten-salt coolant to a third energy conversion loop for generation of electricity or directly hydrogen. If electricity is produced, a multireheat nitrogen or helium-Brayton power cycle is preferred because of the nature of high-average coolant temperature condition of the AHTR. For direct hydrogen production, the high-temperature potential of the reactor outlet coolant temperatures should favor high-temperature steam electrolysis or thermochemical process for efficient hydrogen production. The efficiency at the 1000°C reactor temperature could be 10%–20% higher than at the 700°C reactor temperature depending on the hydrogen production process selected.

Since 2005, R&D has focused on the development of fast-spectrum MSR concepts (MSFR) combining the generic assets of fast-neutron reactors (extended resource utilization and waste minimization) with those relating to molten-salt fluorides as fluid fuel and coolant (low pressure and high boiling temperature and optical transparency). In contrast to most other molten-salt reactors previously studied, the MSFR does not include any solid moderator (usually graphite) in the core. This design choice is motivated by the study of parameters such as feedback coefficient, breeding ratio, graphite life span, and ^{233}U initial inventory. MSFR exhibit large negative temperature and void reactivity coefficients, a unique safety characteristic not found in solid-fuel fast reactors. Compared with solid-fuel reactors, MSFR systems have lower fissile inventories, no radiation damage constraint on attainable fuel burnup, no requirement to fabricate and handle solid fuel, and a homogeneous isotopic composition of fuel in the reactor. These and other characteristics give MSFRs potentially unique capabilities for actinide burning and extending fuel resources.

4.3.6 Supercritical-Water Reactors

The supercritical water-cooled reactor (SCWR) was proposed by the University of Tokyo in 1989 and is a high temperature, high-pressure water-cooled reactor operating above the thermodynamic critical point of water (374°C and 22.1 MPa). It has no flow recirculation system and has a direct cycle primary system. This reactor is an evolutionary development of today's conventional LWRs. SCWRs have an epithermal neutron spectrum but, depending on the core design, may have a thermal or fast-neutron spectrum. Reference power of the reactor is 1700 MW(e). Coolant is heated from 280° C to about 625°C in the reactor core and delivered to a direct power conversion cycle using direct turbine technology. SCWRs offer advantages compared with state-of-the-art LWRs, including (i) high thermal efficiencies of up to 44%; (ii) lower coolant mass flow rate per unit thermal power, offering a reduction in pumping power and reactor equipment; (iii) no risk of boiling crisis (i.e., dry-out) owing to the absence of a second phase in the reactor, thereby avoiding discontinuous heat transfer regimes during normal operation; (iv) simpler design, since steam dryers, steam separators, recirculation pumps, and steam generators are eliminated; and (v) application to hydrogen production and cogeneration. The two main design of SCWR are considered, one with a large reactor pressure vessel with ~0.5 m wall thickness to contain the core and another reactor designs that use distributed pressure tubes analogous to the CANDU reactors. Operating pressures are about 25 MPa with coolant outlet temperatures of 625°C, which facilitates process heat applications. State-of-the-art supercritical steam cycles in conventional power plants are mostly designed with a single steam reheat and regenerative feedwater heating reaching efficiencies of up to 54%. In principle, each steam reheat process increases the thermal efficiency of the cycle. It also reduces the amount of moisture in the last stages of the turbine that must be removed. A single steam reheat raises the efficiency by about 3%–4% compared with no reheat. However, more than two reheat processes are not practicable, since equipment cost would increase with each reheat stage. Therefore, using the high-temperature heat from an SCWR to heat water in the hydrogen production loop is a viable option. Although SCWR was proposed during the early years of the LWR development, it had not attracted significant development interest until new SCWR concepts were emerged in the 1990s in Japan [4], Russia [5], and Canada.

4.3.7 Fusion Reactors

Fusion energy is produced by nuclear fusion of two lighter atomic nuclei to form a heavier nucleus. The latter weighs slightly less than the total of the two nuclei. The small difference in mass m is transformed into energy E according to Albert Einstein's famed formula $E = mc^2$, where c is the speed of light approximating 3×10^8 m/s. Fusion power plants produce heat similar to nuclear fission power plant, which can be used to rotate electric generator to produce electric power. The fusion reaction is induced by bringing two or more light atoms close enough together so that the residual strong force (nuclear force) in their nuclei will pull them together into one larger atom. Fusion between the atoms is opposed by their shared electric charge, specifically the net positive charge of the nuclei. Enough external source of energy must be supplied to overcome this electrostatic force or "Coulomb barrier." The easiest way to do this is to heat the atoms, which has the side effect of stripping the electrons from the atoms and leaving them as bare nuclei. In practice, the nuclei and electrons are left in a high-temperature plasma state so that the nuclei have enough energy to overcome their repulsion. Hydrogen has the smallest nuclear charge, therefore, reacts at the lowest temperature. Most fusion reactions combine isotopes of hydrogen, protium, deuterium, or tritium to form isotopes of helium (3He or 4He). When a nucleus of deuterium fuses with a nucleus of tritium (D-T reaction), an α-particle is produced and a neutron released. The nuclear reaction results in a reduction in total mass and a consequent release of 17.6 MeV per reaction energy in the form of the kinetic energy of the reaction products. Thus, just 1 kg of this fuel would release 3.37×10^{14} J of energy and would provide the requirements of a 3900 MWt power station for a day. This reaction has the highest cross section for fusion at a relatively lower temperature. Compared with fission energy, fusion energy has important benefits. These benefits include abundant supply of fuel (hydrogen isotopes), less radioactivity on the products, and inherently safe as there is no critical mass. Further fusion source being inherently of high temperature enables variable and economical energy products including hydrogen from water. The current leading designs of confining plasma in the fusion reactor are mainly two type, the magnetic confinement such as in tokamak and inertial confinement (ICF) by laser. The International Thermonuclear Experimental Reactor (ITER) tokamak in France and the National Ignition Facility laser in the United States are two major programs in these technologies. Fig. 4.12 shows the ITER fusion reactor concept for power production, where (1) D-T fuel injected (2) reaction in confined plasms and (3) reaction products including helium and fast particles or radiance (4) heat deposited breeding blanket, (5) heat transfer, and power conversion. The heat is used to produce steam and supply a conventional turbine and alternator electricity producing system (5).

Fusion power has a potential to produce the high-temperature heat in the blanket, and it can be applied to the hydrogen production. Various blanket concepts have been studied to extract the heat from fusion. Such one design is the advanced helium-cooled pebble-bed blanket of European demonstration fusion reactor (DEMO) blanket concepts, made from SiC_f/SiC composite as structural material. Lithium ceramic such as Li_4SiO_4 in pebbles with a typical diameter of 0.4–0.6 mm is used as breeding material, and beryllium or beryllide in pebbles with a diameter in the range of 1 mm is used as neutron multiplier material.

Fusion reactor heat and electric power can be used for HTSE process and thermochemical water-splitting SI cycle to produce hydrogen. About 1000° C temperature heat obtained from this high-temperature blanket can drive thermochemical SI cycle. However, some important technical issues need

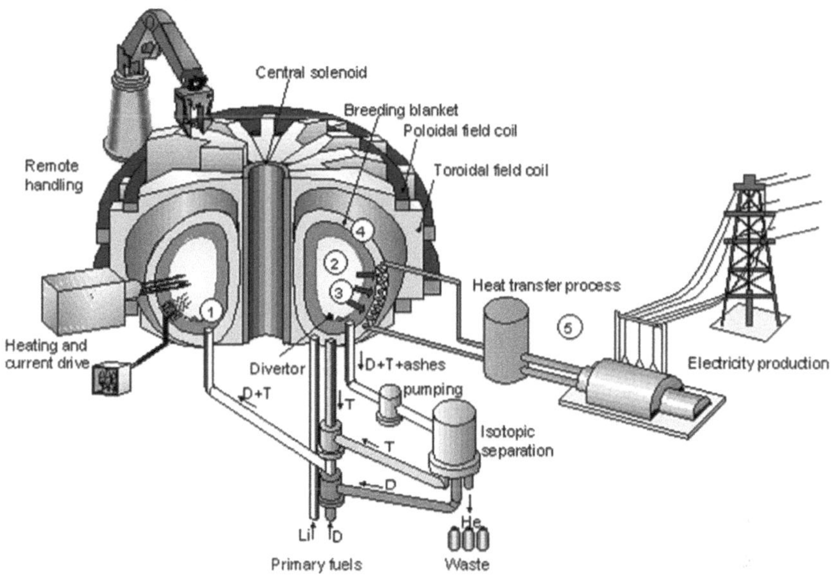

Fig. 4.12 Tokamak fusion reactor power plant.

further development, including material selections for hydrogen production system and intermediate heat exchanger, intercontamination issues between hydrogen production system and fusion reactor, and R&D on the subsystems like power conversion system and tritium control system.

4.4 NUCLEAR HYDROGEN TECHNOLOGY

As discussed earlier, hydrogen can be produced using nuclear reactors either by electrolysis or by thermochemical processes without any contribution of CO_2 to atmosphere. The current operating light-water or heavy-water reactors and gas-cooled reactors are limited in maximum temperature of 350°C. So, these reactors can be used for electrolysis at low temperature. For low-temperature electrolysis process, the requirements for additional infrastructure are minimum, because electric power lines can be used at any distance from the nuclear plant and electrolysis plant can be located at the demand site. Other than economic issues, the nuclear hydrogen technology does not have any major issues for hydrogen production. High-temperature steam electrolysis requires temperatures in the range of 700–1000°C. Current LWRs and near-term, water-cooled advanced LWRs produce temperatures under 350°C and cannot be used for such purposes. However, other coolants of several Gen IV reactor concepts are proposed to reach such high temperatures (above 700°C) and can be coupled to thermochemical plants. Table 4.6 shows nuclear hydrogen production options.

Nuclear hydrogen production processes such as SMR, thermochemical water decomposition, and high-temperature steam electrolysis require a wide range of thermal and chemical reactors and other equipment. The hydrogen production plant components should maintain structural integrity at service temperature and pressure that can approach to 900°C and 4 MPa, heat and resistance to corrosive process fluids, affordable material and fabrication cost, and sufficient lifetime. Although SMR is established and widely practiced process industry, it requires new safety-related design and operational consideration for its coupling to nuclear plant. There are number of specific safety and design consideration required for nuclear hydrogen production plant. In an HTGR-SMR hydrogen production system, the pressure difference between the secondary helium and process gases is an important parameter. The reformer reaction tube constitutes pressure boundary between the heat transport secondary helium gas from HTGR and the process gas of a mixture of methane and steam. The reformer reaction tube wall thickness should be able to withstand maximum pressure difference between helium and process gases in not only during normal operations including start-up and shutdown but also during off-normal conditions such as a reduced or loss of chemical reaction.

4.4.1 Nuclear Hydrogen System Integration

Hydrogen production systems coupled with a nuclear reactor introduce new considerations and requirements in the design and operation of coupled nuclear hydrogen plant. The nuclear plant and hydrogen generation plant should be colocated to exchange the heat energy. The interface between the reactor and hydrogen production system involves potentially long heat transfer paths at high temperatures, heat exchangers that are subject to both high-temperature and corrosive chemical environments, new safety and regulatory issues, and infrastructures for chemical processes and hydrogen and oxygen transfer and storage. These issues will be common to any nuclear hydrogen plant, however many more issue will depend on the specific hydrogen production process. The coupled nuclear hydrogen plant interface and balance of plant are shown schematically in Fig. 4.13. The high-temperature reactor provides nuclear process heat to the hydrogen plant

Table 4.6 Nuclear Hydrogen Production Options

	Hydrogen Production Method			
	Electrolysis		**Thermochemical**	
Feature	**Water**	**High-Temperature Steam**	**SMR**	**Water Splitting**
Required temperature (°C)	>0	>300 for LWR >600 for Cu-Cl cycle	>700	>850 for SI cycle >600 for S-AGR
Efficiency (%) of chemical process	75–80	85–90	70–80	>45, depending on temperature
Efficiency (%) coupled to LWR	27	30	Not feasible	Not feasible
Efficiency (%) coupled to HTGR, AHTR, or S-AGR	Below 40	40–60, depending on temperature	>70	40–60, depending on cycle and temperature
Advantages	Proved technology with LWRs	Can be coupled to reactors operating at intermediate temperatures	Proved chemical process	Eliminates CO_2 emissions
			40% reduction in CO_2 emissions	
	Eliminates CO_2 emissions	Eliminates CO_2 emissions		
Disadvantages	Low efficiency	Requires high-temperature reactors	CO_2 emissions are not eliminated	Aggressive chemistry
		Also requires development of durable HTES units	Depends on methane prices	Requires development

LWR, light-water reactor; *S-AGR*, supercritical CO_2 advanced gas reactor; *SI*, sulfur-iodine; *Cu-Cl*, copper-chlorine; *HTGR*, high-temperature gas-cooled reactor; *AHTR*, advanced high-temperature reactor; *HTES*, high-temperature electrolysis of steam.

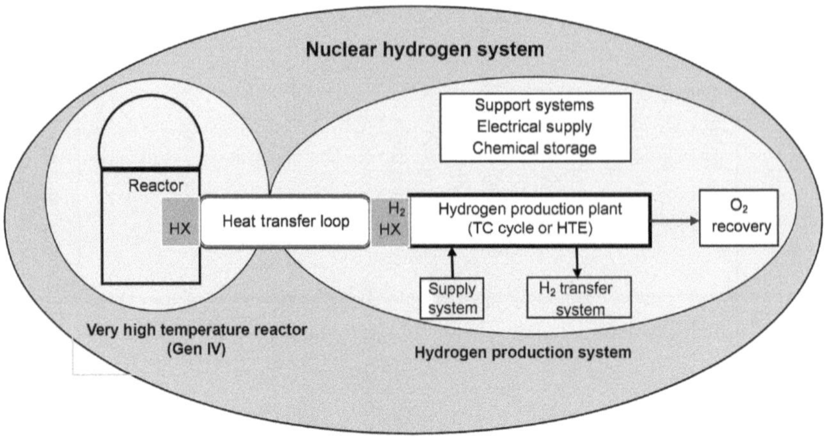

Fig. 4.13 Schematic of the interface between nuclear reactor and hydrogen plant and balance of plant.

through an intermediate heat transfer loop. The reactor may also provide electric power to the hydrogen production plant. The intermediate loop working fluid and materials of construction are important engineering issues at these elevated temperatures. For safety purpose physical isolation, either separation distance or an acceptable physical barrier should be provided to eliminate the propagation of accident consequences from one plant to the other. Additionally, numerous supporting systems are required for gas handling, storage, and process support. Special high-temperature heat exchangers are required to transfer heat from the reactor primary systems to the hydrogen process for efficient coupling of the nuclear hydrogen plant. Table 4.7 list the issues on balance of plant and plant interface between nuclear systems [6]. Key considerations include the development of the high-temperature interface components (the heat exchangers and intermediate loop) and materials required to couple the thermochemical or high-temperature electrolysis system to the high-temperature heat source. Interface requirements are defined by the characteristics of the thermochemical and HTE processes and on developing the components needed to demonstrate selected processes.

For integration of the nuclear and hydrogen plant, design studies should address number of issues. The design studies for the system interface and balance of plant areas should address plant configuration options for both thermochemical and high-temperature electrolysis systems, operation and control, safety and isolation issues for the coupled plants, and assessment

Table 4.7 Considerations on Nuclear Hydrogen Plant Interface and Balance of Plant

Systems/Components	Design/Interface Considerations
High-temperature heat exchangers (HX)	• HX interface to production process, HX design options, operational conditions (temperature, pressure, and fluids) • Materials compatible with high-temperature, heat transfer medium and process chemical species
Intermediate loop	
Heat transfer medium	• Heat transfer conditions—temperature, pressure, pumping power, heat loss requirements, working fluids
High-temperature transfer lines	• Isolation configuration, materials, operating conditions (temperature, pressure, and fluids), insulation, seals, auxiliary heating
Support systems	
Oxygen recovery/disposal oxygen safety systems	• Oxygen inventory; recovery; or disposal approach, storage, safety issues • Oxygen hazards (combustion and health), material corrosion, ventilation, fire systems, diagnostics, environmental control
Hydrogen safety systems	• H_2 diagnostics, storage, combustion, ventilation systems, explosive hazards
Hydrogen transfer chemical support system instrumentation and control electric alternative heat source	• Transfer system to interface with pipeline, storage systems • Chemical feed, storage, purification, diagnostics, process controls • H_2 process diagnostics and controls, interface with reactor systems • Hybrid process, grid, station power systems • Interim process heat requirements for testing or production

of applicable codes and standards. Hydrogen plant configuration studies should define various configuration options and the operational conditions and requirements for the hydrogen plant subsystems. The isolation methods for nuclear and hydrogen production systems, both thermochemical and electrolytic processes, have performance and economic implications. These studies define options and trade-offs for the optimum coupling of the nuclear and chemical systems. System interaction studies should address safety and isolation issues arising from system level considerations and from functional and physical coupling of the nuclear and hydrogen plants. Nuclear hydrogen production systems will require a new or modified framework for regulation and applicable code and standards.

High-temperature heat exchangers are critical components of the nuclear hydrogen plants. The heat exchanger design should be able to handle elevated temperatures and pressure, and chemicals that may be highly corrosive such as heat exchanger for SI cycle require decomposition of sulfuric acid at more than 900°C and up to several megapascals (MPa). The heat exchanger for the HTSE will be a steam generator producing steam at up to 950°C and up to 5 MPa. Intermediate heat transfer loop interfaces with the reactor heat source where the heat transfer media, line configuration, and material are key technology issues. The selection of the medium/fluid for the intermediate heat transfer loop has to be chosen to optimize plant configuration, separation distance decisions, heat exchanger design, and material selection. The medium should be compatible with high temperatures up to 1000°C, pressures of several MPa, and reasonable pumping and circulating requirements, as well as chemical compatibility with heat transfer loop materials. Fluoride molten salts, helium, and liquid metal are good candidates for the heat transfer medium. These media meet the high-temperature and low-pressure requirements. Fluoride salts have viscosities similar to water that translate to relatively low pumping costs. Liquid metals are expensive systems, have purity and pumping component issues, and have high densities and viscosities that will result in higher pumping costs over long distances. Helium as a gas needs large volume of medium resulting in large line sizes and large blowers.

The coupling configuration of nuclear power plant (NPP) and hydrogen production plant (HPP) depends on the reactor type and the method of hydrogen production. The NPP can be connected to a power conversion system such as direct Brayton cycle to produce electricity that feeds power to hydrogen plant such as in for HTSE process as shown in Fig. 4.14A. An example of high-temperature reactor directly coupled to SI cycle is shown in Fig. 4.14B. The reactor can be connected to hydrogen plant with an intermediate heat exchanger (IHX) or pair of series IHX through various configuration including the following options:

1. NPP directly connected to HPP with IHX or series of IHX for thermochemical hydrogen generation
2. NPP connected to direct Brayton in series with IHX for HTSE

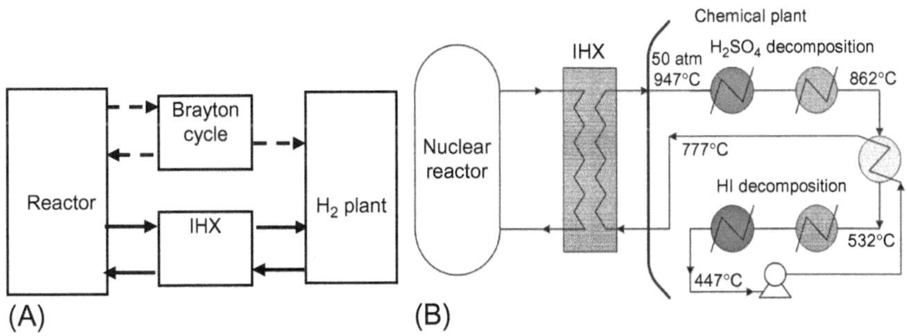

Fig. 4.14 (A) Coupling of high-temperature nuclear plant to hydrogen plant and (B) schematic diagram of VHTR connected to SI cycle-based hydrogen plant.

3. NPP directly connected to HPP with IHX and NPP also connected to direct Brayton cycle in series with IHX for hybrid process
4. NPP connected to indirect Rankine cycle in series with IHX indirectly coupled to HPP for HTSE
5. NPP connected to HPP with IHX and to direct Brayton cycle for cogeneration
6. Separate nuclear plants for thermal coupling and electricity generation

4.4.2 System Safety

Safety is very important particularly when nuclear technology is involved due to radioactive material present in the system. The safety objective with nuclear power plant operation is to protect people and the environment from harmful effects of ionizing radiation. Nuclear reactor accidents such as Chernobyl, Three Mile Island, and Fukushima have left the public concerned about the safety of nuclear power in commercial use. Advanced new light- and heavy-water reactors and Gen IV nuclear plants have a strong emphasis on passive and inherent safety features. In nuclear hydrogen system, the presence of hydrogen as a highly explosive gas presents even more safety considerations. The safety design of nuclear power plant is based on the defense in depth philosophy. All nuclear power plants are designed with dedicated safety systems to prevent accidents and mitigate their consequences. Thorough safety design review of the safety system is carried out to prevent accidents and to mitigate their consequences. Basic safety functions of the nuclear power plant are to control reactivity in the core, to remove heat from the core, and to confine radioactive materials released from the primary coolant and the core. The safety system maintains their safety functions against internal and external events. For nuclear hydrogen plants, external events include fire, explosion, asphyxia, release of toxic gases and corrosive gases and liquids that originate in the hydrogen plant.

The reactor core, control, and safety systems are designed to withstand static and dynamic loading in normal operation and accident to ensure the safe shutdown of the reactor, to maintain the reactor at subcritical condition, and to adequately cool the core. Reactivity control system has redundancy, diversity, and independency to avoid common mode failure and reduce probability of failure. In case of abnormal behavior, the reactor is shut down with control rods pushed into the core, and the cooling system removes the residual heat. In an accident scenario, with loss of cooling, emergency core cooling system provides adequate core cooling. The Gen IV reactors such as VHTR and SFR have inherent safety design; in case of overheat, the reactor automatically shuts down due to negative temperature coefficient of reactivity. For long-term decay, heat removal by passive mode of cooling is provided such as thermal conduction and natural circulation so that no external power or interrogation is required. In coupled nuclear hydrogen plant, the reactor coolant system including the intermediate heat exchanger is designed to withstand the static and dynamic loading in all operational states including accident states to prevent the release of radioactive materials with the coolant.

A reactor containment system provides a strong barrier to prevent the amount of radioactive material release into the environment in an accident. The reactor containment is a leak-tight large volume retaining radioactive materials released from the primary coolant system and withstands internal pressure rising with the discharge of coolant to ensure the structural integrity. The amount of coolant released into the environment from the stacks in the design basis event is acceptable. The reactor core and containment is protected from the air leakage from the outside of containment to avoid any oxidation. The heat transfer loop that connects the nuclear plant with the hydrogen production plant to transfer hot helium gas for hydrogen production is designed with isolation valves that close to isolate reactor in case of tube failure or breach of the line outside the containment.

A hydrogen plant connected to the nuclear plant contains large amount of combustible gas and toxic gas. An accidental release of hydrogen gas is one of the most important external events in the nuclear hydrogen production plant as there is the risk from fire and explosion. Hydrogen being light can travel easily to distance from breach area and can cause fire and explosion hazard. If ignited during leakage, jet flames are formed that may damage components by overheating. The reactor plant and hydrogen plant should be located with adequate separation or have a physical barrier between them to avoid fire and explosion hazards. Similar to chemical plants, leak detectors and emergency shutoff valves should be provided for detecting and stopping a leakage of hydrogen. The length of the jet flame may be several meters, and the safety items in the nuclear plant should be placed at hundreds of meters away from the hydrogen production system, so a jet flame would not directly damage any nuclear safety-related systems. In case of a hydrogen-air gas cloud explosion, the resulting overpressure may damage the reactor building or components installed outside the nuclear plant. Tritium is generated inside the reactor by ternary fission of the fuel due to thermal neutrons and neutron capture reactions of lithium included in graphite components, boron in graphite and control rods, and helium-3 in coolant in VHTR. The tritium may transport from the reactor core to SI cycle through the IHX because of high diffusivity and results in migration of the radioactive material into product hydrogen. The intercontamination about tritium and hydrogen between fusion reactor and hydrogen process can be reduced with additional intermediate helium loop separating them. Isolation valves are a key safety component when coupling a nuclear reactor to a hydrogen production plant. In the case of a pipe rupture, it is essential to isolate the primary loop to prevent the release of helium contaminated by fission products.

The key requirements for coupling a hydrogen production plant with nuclear plant are (i) the assurance of the safety of the nuclear power plant against postulated events initiated in the hydrogen plant in order to guarantee cooling of secondary helium circuit during normal operation and maintaining the differential pressure between primary and secondary helium circuit and (ii) the construction and operation of the hydrogen production plant as a conventional, nonnuclear facility by mitigating tritium concentrations in the hydrogen plant below the limits allowed by regulation in order to guarantee a radioactivity-free hydrogen plant.

The hydrogen production plant is an energy sink in nuclear hydrogen system. The systems are energetically coupled and hence dynamically affect each other during operation. A transient can be initiated on either the nuclear reactor side or the chemical side of the plant that can affect both. There are many potential transients that occur in a nuclear reactor system. Some examples are start-up or shutdown, reactivity insertion or removal, off-normal operation, and design basis accident. In the chemical plant, examples of transient driving forces are reaction chamber temperature change, small pipe break or leak, vapor explosion, and chemical plant fire or another event. In terms of nuclear reactor response, each of these chemical plant events is a loss-of-heat-sink accident. When unifying two systems that are dynamic and provide feedback to each other, the nature of the response is dictated by the relative time constants of the plants. For example, consider a VHTR connected to SI cycle. For the VHTR is a thermal reactor, the delayed neutrons are the important factor in reactor response. A thermal reactor has a time constant of about 55 s. In the chemical plant, depending on the process, the response time can be small or large, for example, in SI cycle. The reaction time constant for sulfuric acid decomposition is in the order of 20 s, whereas the HI decomposition has a response time on the order of 500 s. The limiting reaction rate in the chemical plant is that of HI decomposition. Since the chemical plant is composed of cyclic processes, the slowest reaction rate provides a first-order approximation of the plant response. A malfunction of equipment or control systems in the SI cycle can induce temperature fluctuation of helium at the outlet of the SI cycle because of the variation of heat load. Several dynamic models of hydrogen generation plants coupled to nuclear reactors have been presented in literature. A simulation study performed with a pebble-bed modular reactor (PBMR-268) coupled to SI cycle similar to the system shown in Fig. 4.14 with PBMR kinetic and thermal-hydraulic models, and chemical kinetics and reaction models for SI cycle, indicates that the effects of overcooling can increase the fuel temperature [7]. In Fig. 4.15, for the overcooling of helium inlet temperature by 150 K for 300 s period, the transient of fuel temperature and reactor relative power is shown.

Overall, the potential risk of public should be low enough when a hydrogen production plant is coupled with nuclear plant. It is considered that a nuclear hydrogen production plant can maintain the safety function of a nuclear plant even though the hydrogen production plant is in abnormal state such as thermal disturbance, combustible gas release, and other possible transients.

4.4.3 Licensing Considerations

Licensing of the nuclear plants being an important aspect of regulation, it is carried out with stringent safety criteria being set and met. Before a commercial nuclear reactor can be built and operated, approval is usually obtained from regulatory authorities as designated by the government. The regulatory body is responsible by law for licensing and regulating nuclear reactors and

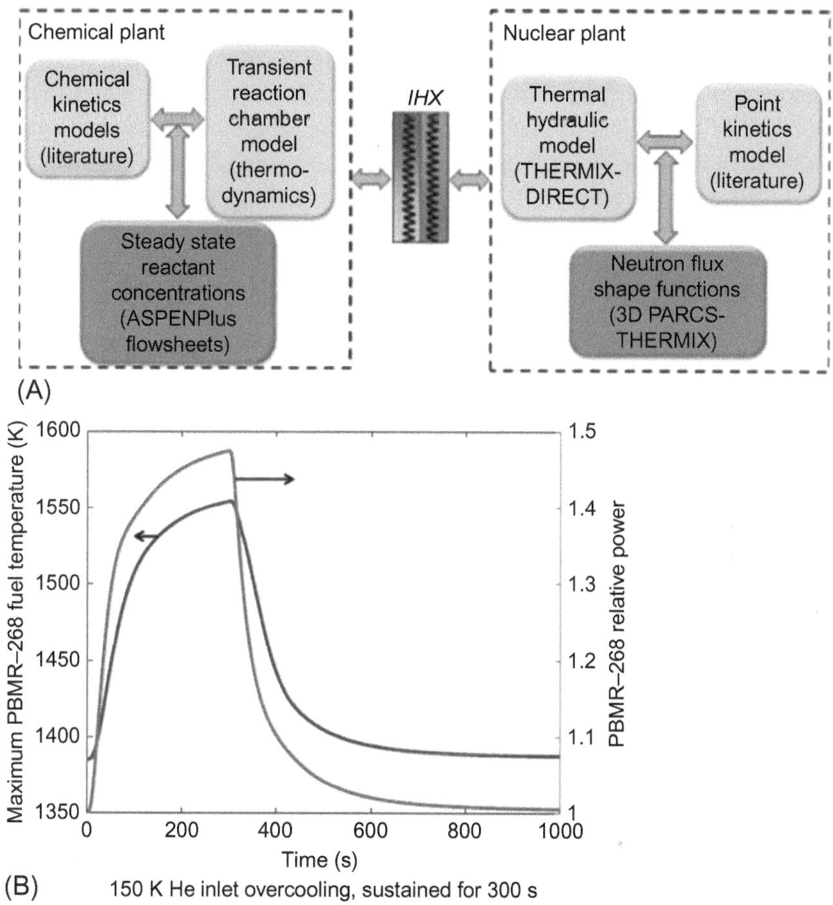

Fig. 4.15 (A) Models for a PBMR-268 coupled to a SI cycle hydrogen generation plant. Furthermore, the intercontamination, (B) PBMR maximum fuel temperature, and power for overcooling of coolant.

materials for their intended safe and peaceful purposes while protecting the safety of people and the environment. A nuclear hydrogen production plant will require nuclear regulatory approval. In the United States, the Nuclear Regulatory Commission (NRC) has statutory responsibility for licensing and regulating the construction and operations of commercial power reactors including the LWRs and developmental reactors. Currently, operating nuclear power plants have been licensed under a two-step process described in Title 10 of the Code of Federal Regulations (10 CFR) under Part 50. This process requires both a construction permit and an operating license. Since 2009, the US NRC issues combined construction and operating license

(COL) governed by Title 10 of the Code of Federal Regulations (CFR) under Part 52, simply known as 10CFR52. In addition under Part 52, licensing options include early site permits that allow an applicant to obtain approval for a reactor site without specifying the design of the reactor(s) that could be built there and certified standard plant designs that can be used as preapproved designs. Similarly, other countries have licensing process that is essentially a major component as in US NRC licensing guidelines. All licensing processes involve detailed design review, safety analysis, environmental review, and antitrust review.

To construct or operate a nuclear power plant, an applicant must submit a safety analysis report (SAR) that contains the design information and criteria for the proposed reactor and comprehensive data on the proposed site. The SAR discusses hypothetical accident situations and the safety features of the plant that would prevent accidents or lessen their effects. In addition, the application must contain a comprehensive assessment of the environmental impact of the proposed plant. A prospective licensee also must submit information for antitrust reviews of the proposed plant.

The licensing process under COL is shown in Fig. 4.16 where safety review, environmental review, and public involvement and hearing are performed before license is approved.

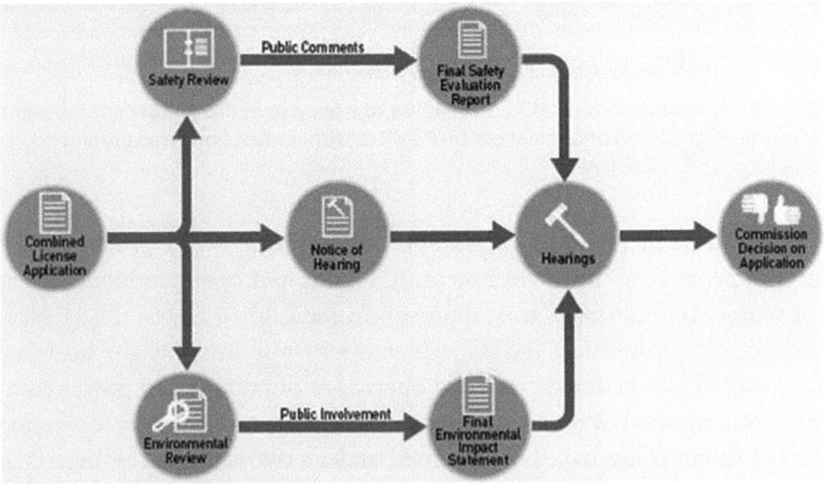

Fig. 4.16 US NRC licensing process under combined license. *(US NRC Nuclear Power Plant Licensing Process, US Nuclear Regulatory Commission Report NUREG/BR-0298, July 2009.)*

For the new-generation reactor such as VHTR, licensing framework is being developed. Several test and demonstration high-temperature gas-cooled reactors have been built and operated in the United Kingdom, Germany, the United States, Japan, and China, of which two test reactors are in operation today and their licensing practices are applicable to future HTGRs.

The high-temperature engineering test reactor (HTTR) is an operating test reactor for the HTGR in Japan. The HTTR has inherent safety features such as strong negative reactivity feedback coefficient, large heat capacity of the core, and inert helium gas as coolant and relies on passive decay heat removal. The safety evaluation procedure of the HTTR is also different from that of current reactors. The Japan Atomic Energy Research Institute (JAERI) established safety criteria for the fuel, core structures, and high-temperature components. The acceptance criteria for anticipated operational occurrences of HTTR are the following: (i) The maximum fuel temperature shall be <1600°C to avoid fuel failure; (ii) pressure on reactor pressure boundary is <1.1 times of maximum pressure in service; and (iii) maximum temperature of reactor pressure boundary for 9/4Cr-1 Mo steel <500°C, austenitic stainless steel <600°C, and Hastelloy XR < 980°C during the operational transients. Acceptance criteria for the major accident and the hypothetical accident are the same as that for the LWRs. Similar efforts are underway at US NRC on developing framework for licensing of new-generation nuclear plant that include design-specific COL strategy, development of technical approach to establishing the NGNP licensing basis and requirements, and development of analytic tools. The major technical areas for new-generation reactors include accident analysis, fuel performance and fission product transport, high-temperature materials and graphite performance, and process heat applications.

The key issues of the licensing approach of the nuclear hydrogen plant include the following: (1) definition of design criteria of fuel, graphite, and high-temperature components; (2) technical basis for the use of a confinement; (3) definition of source term; and (4) combined plant safety. For the future commercial VHTR plants, it is necessary to develop fuel to meet the high power density, high burnup, and high outlet gas temperature. The qualification of coated fuel particle technology is an important development task for the licensing. Simultaneously, the standards of quality assurance and inspection should be developed. As graphite will be an essential material for HTGR core structural and fuel components, qualification of different grades of graphite is important. Qualification of materials for the high-temperature and extended lifetime conditions is critical, and alternative materials must be carefully evaluated. It is necessary to ensure the design of confinement that the fuel performance model is developed; the fission product released to the environment and the public dose are calculated and identified to satisfy the criteria. The assessment of radiological consequences for licensing-basis events on the basis of event-specific mechanistic source terms needs to be considered. Finally, the safety of the coupled systems has to be established. For reactor safety analysis, the hydrogen plant aspects may be considered in the same category as external events.

4.4.4 Nuclear Hydrogen Plant Economics

The economics of nuclear hydrogen production depends on a wide variety of parameters such as nuclear plant size and plant availability, cost of feedstock, efficiency of the technology employed, maturity of the technology, and physical distance to end-use markets. Though several cost analyses for hydrogen can be found in the existing literature, they are not always comparable. Currently, economies of scale clearly apply to large SMR plants with a clear operating cost advantage over smaller units designed for distributed hydrogen production applications. The price of natural gas plays key role on the cost of SMR hydrogen. Hydrogen production costs are primarily determined by the investment costs. The cost estimate is influenced by the production method and the amount of the production. The share of fuel cost, for example, is in the order of 50%–68% in large-scale SMRs, whereas it is 28%–40% in small reformer units. The cost of electrolytic hydrogen is dominated by the electricity cost (75%–80%). In biomass gasification, fuel costs are about 40%. The fossil primary energy cost varies with time and thus affects hydrogen production cost.

It is known that that the nuclear is one of the least expensive low-carbon energy sources and is less vulnerable to fuel price changes than are the fossil energy sources. Advanced nuclear driven thermochemical processes are still under development for commercial units; however, they may provide low per unit production costs in the future due to steady and lower nuclear fuel costs. Moreover, the product benefits from oxygen sales. For nuclear plant, operating and maintenance costs include decommissioning costs in addition to the usual labor, taxes, security, and other costs. There are several options where the nuclear power can be taken advantages of with changing prices of electricity and heat. One of the options with nuclear hydrogen generation is that during low electricity demand, electrolytic hydrogen generated from nuclear power can be stored and then reconverted to electricity in times of peak demand. Similarly in heating case, the steam price increases with increasing natural gas prices, while the price of steam produced by nuclear power remains constant as gas prices go up. This can make the price of steam production from nuclear power competitive. High-temperature reactors offer higher thermal-to-electric conversion efficiencies and with high-temperature electrolysis providing steam at the 800°C, when HTSE is operated in the reverse mode as a fuel cell produces electricity with same equipment reducing the capital cost investment. In addition to the fuel cell plants, highly efficient hydrogen-oxygen gas turbines could be operated to meet peak demand. For better economics, the system requires inexpensive large-scale storage of both hydrogen and oxygen such as underground storage in geologic formations with sufficiently low permeation rates similar to today's storage of natural gas. Transportation of the hydrogen still appears to be a comparatively expensive part of the cost of delivered hydrogen, which can be minimized by focusing on centralized production, storage, and use of the hydrogen for peak electricity production. Levelized hydrogen production cost estimate performed by ANL for three nuclear hydrogen systems, ALWR and high-pressure low-temperature electrolysis HPE, HTGR and SI cycle, and HTGR and HTSE, showed that HTGR-SI is more economic with \$2.26 per kg of hydrogen versus HTGR-HTSE at \$2.51 and ALWR-HPE at \$2.91 [8]. However, there is uncertainty in future prices, and the flexibility to switch between hydrogen and electricity leads to significantly different results in regard to the relative profitability of the different technologies and configurations.

There have been wide range of efforts in several countries on the projected economic potential of hydrogen production and cost assessments to direct and prioritize research and development efforts and to compare and select the most appropriate processes. Several models for technology and economic assessment have been developed such as the German-French E3-database, the hydrogen cost analysis model in Canada, and H_2A modeling tool developed by the United States Department of Energy (USDOE) Office of Energy Efficiency and Renewable Energy, and other models in the United States. Under international collaboration on methodology for estimating costs of nonelectricity products, nuclear energy has been developed by GIF Economics Modeling Working Group. The methodology separates costs into capital costs and annual costs. The generation IV excel calculation of nuclear systems (G4-ECONS) software package is an "integrated nuclear energy economic model" consisting of several nuclear-economic submodels. G4-ECONS includes models for capital investment, production costs, nuclear fuel cycle, operations and maintenance (O&M), optimal scale, and energy products. Construction costs are integrated into capital costs, which are borrowed and amortized over a fixed time period. The annual costs, which are recurring over the lifetime of the system, include O&M costs. The working group also suggests the use of the "power credit method" for costs that may be associated with dual product and electricity generation. This method has been adopted by the IAEA to evaluate the economics of nuclear desalination. The power credit method first calculates total expenses and energy from a single-purpose electricity producing reactor and derives a cost per kilowatt hour. Then, the amount of nonelectric product and energy and the total expenses of the dual-purpose plant are calculated. Electricity production from a dual-purpose reactor is lower than for a single-purpose reactor, but the costs for the dual-purpose plant are higher.

The IAEA has developed the Hydrogen Economic Evaluation Program (HEEP) to assist member states in performing economic analyses for nuclear hydrogen production processes. The most promising hydrogen production technologies including conventional and high-temperature electrolysis and thermochemical and hybrid thermochemical cycles such as SI cycle and HyS cycle are integrated with different nuclear power plants in the HEEP. HEEP is capable of performing comparative studies with different nuclear energy sources for hydrogen production and for electric power cogeneration. The package also facilitates performing a broader investigation for more understanding of the feasibility of different hydrogen production scenarios including storage, transportation, and distribution, with user options. It consists of modules on nuclear plant, hydrogen plant, hydrogen storage, and transportation. The cost element is associated with each modules. Fig. 4.17 shows the four modules in HEEP. The cost elements of the modules are shown in Fig. 4.18. HEEP estimates the levelized cost of hydrogen, considering many

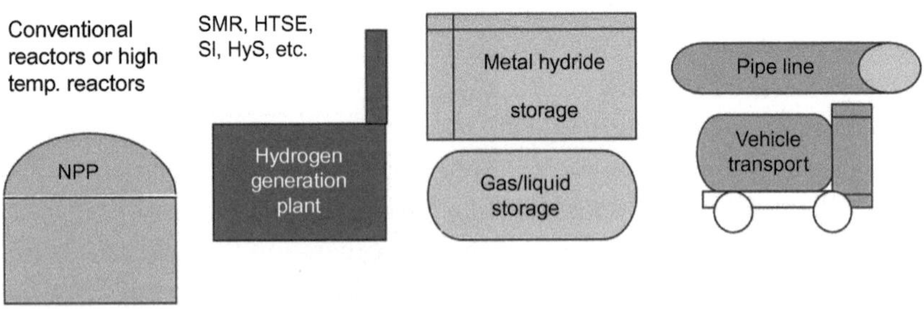

Fig. 4.17 The HEEP modules for nuclear plant and hydrogen generation, storage, and transport.

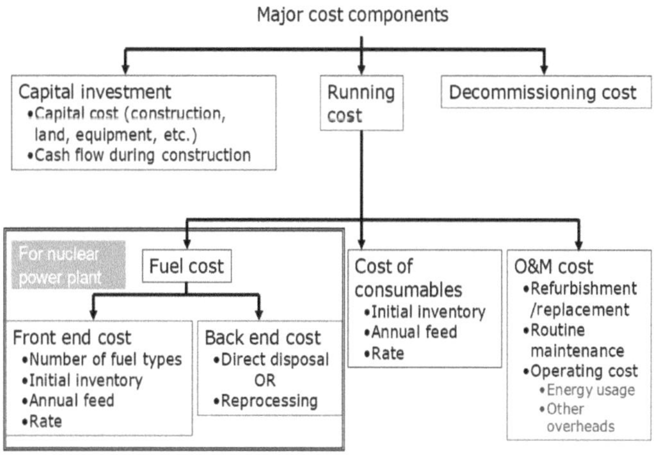

Fig. 4.18 Important cost components considered in HEEP.

aspects of capital investments that eventually affect the calculated cost. The capital investment, which is the sum of all expenditures incurred in design, licensing, manufacturing and erection, construction, and commissioning of the plant, can be raised at a given equity to debt ratio, that is, the funding of the project can be raised through equities or can be raised through market borrowings or combination of both. The cash flow during construction is also another important parameter, which affects the interest during construction. This is particularly prominent in the case of nuclear power plant requiring high capital investments with longer construction period. The updated version of HEEP was released in December 2014 and is available on the IAEA website. Recently, HEEP benchmarking with H_2A and G4-ECONS software was conducted by IAEA with 10-member state representatives. In Fig. 4.19, a levelized hydrogen cost estimates from HEEP is compared with H_2A and G4-ECONS. This case considered SCWR as nuclear plant with the capacity of 1200 MWe coupled with HTSE plant for hydrogen generation with capacity of 1925 Mm^3/year. Capacity factor of 90%, operational period of 40 years, and discount rate of 5% were considered.

4.5 WORLDWIDE NUCLEAR HYDROGEN R&D

4.5.1 Argentina

A government research and development (R&D) institution, Comisión Nacional de Energía Atómica (CNEA) of Argentina, is responsible for

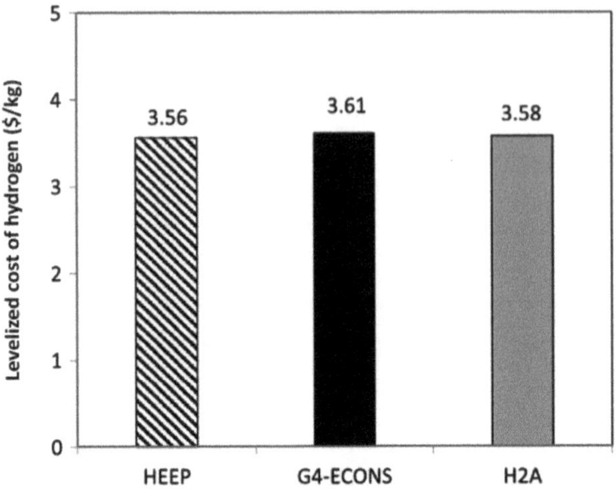

Fig. 4.19 Levelized hydrogen production cost comparison for SCWR coupled to HTSE process.

nuclear policy and overseas R&D program on nuclear energy, applications, safety and environmental protection, and nonnuclear research and applications. In 2006, Argentine Congress passed the Hydrogen Law that promotes and regulates the use of hydrogen in industry, transport, and electric power stations, with the objective of diversifying the current energy matrix of the country, which is largely based on petroleum oil and natural gas. The research activities on nuclear hydrogen currently underway in Argentina are focused on two processes: (1) coal gasification using carbon dioxide and steam and (2) metallic chlorides water-splitting thermochemical cycles. Experimental studies have been conducted on the gasification of subbituminous Río Turbio coal. This coal is very reactive to gasifying agents with gasification reactions occurring at low temperatures, and with a fast kinetics, the hydrogen can be easily separated from the synthesis gas with a double-layer ceramic separating membrane. The thermochemical process-based nuclear hydrogen project is focused on chlorine cycles for hydrogen production, which are one of the leading long-term methods including vanadium-chlorine cycles, cobalt-chlorine cycles, iron chlorides, and coupled iron-copper chlorides cycles [9]. These metallic chloride cycle studies have been advanced on the kinetics and mechanisms of thermochemical reactions at laboratory scale, in order to find the optimum conditions for increasing the efficiency of these cycles with the objective of a future scaling up of the experimental facilities.

On hydrogen storage and transportation of hydrogen, research activities in CNEA are focused on the use of metallic hydrides with a high efficiency in absorption/desorption processes that are being manufactured. The CNEA is studying two types of fuel cells, polymer electrolyte membrane (PEM) and solid oxide fuel cells (SOFC), for the generation of energy with hydrogen. The R&D work on SOFC is on the development of new materials for the cathode produced with innovative methods that allow modifying their microstructure. The work on PEM fuel cell is on the development of better membranes and the manufacturing of carbon-supported nanophase catalysts using three methods of deposition.

4.5.2 Canada

Canada blessed with significant natural resources is one of the world's largest energy producers (ranking fifth) and exporters of energy. Almost all of Canada's energy exports go to the United States. In 2010, Canada produced mostly crude oil (29%), natural gas (28%), and refined petroleum products (22%); the rest was coal (7%) and electricity from primary sources (8%)—hydro, nuclear, wind, and tidal. Hydrogeneration, the largest primary source in 2010, accounted for 63% of electric power and totaled 346.7 million megawatt hours. Oil sand project has increased Canadian oil production, but it has been sharply criticized for its impact on the environment and its intensive use of both water and natural gas. Canada has a matured nuclear industry with more than 60-year evolutionary development of CANDU power reactors built and successfully operated in Canada and elsewhere. Currently, about 15% of Canada's electricity comes from nuclear power, with 19 reactors mostly in Ontario province providing 13.5 GWe of power capacity. Annual hydrogen production in Canada currently is 3 million tons of which 35% is used for chemical products, 24% for refining, 23% for heavy oil upgrading, and 18% for chemical by-products. Canadian hydrogen R&D activities have been active since 1985 within the Canadian Hydrogen and Fuel Cell Program managed by Natural Resources Canada. World's first demonstration of a fuel cell bus in 1993 based on Ballard's polymer electrolyte membrane (PEM) fuel cells and the Hydrogenics low-temperature water electrolyzer are some of the successes in this technology. Recently, thermochemical water decomposition is an emerging technology for large-scale production of hydrogen.

Among various cycles, Cu-Cl cycle has a significant advantage over these other cycles, due to lower-temperature requirements around 530°C and lower. This temperature range match very well with the envisaged Canadian Gen IV SCWR that has great potential to produce high-pressure steam at around 625°C and 25 MPa. The steam at this condition is also very desirable in many petrochemical operations (e.g., traditional oil and oil sand exploitation and bitumen upgrading) and other industrial applications. Several advantages of the Cu-Cl cycle over others include lower operating temperatures, relatively low voltage required for the electrochemical step, ability to utilize low-grade waste heat to improve energy efficiency, and potentially lower cost materials. The University of Ontario Institute of Technology (UOIT), Atomic Energy of Canada Limited (AECL), Argonne National Laboratory (ANL), and other partner institutions have been collaborating on the development of enabling technologies for the Cu-Cl cycle, through the Generation IV International Forum (GIF). The main four steps of Cu-Cl cycle method developed by this collaborative work are shown in Table 4.2. The schematic of the Cu-Cl-based hydrogen production scheme is shown in Fig. 4.20.

The use of the Canadian Advanced CANDU Reactor (ACR) for hydrogen production is considered a realistic short-term approach where using electricity and heat (to produce hydrogen and steam, respectively) would

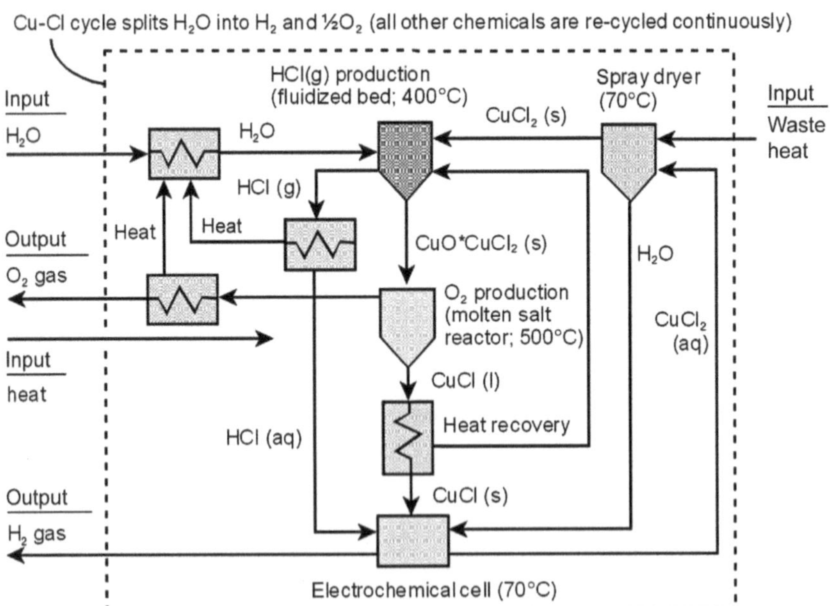

Fig. 4.20 Cu-Cl-based hydrogen production scheme.

represent a first step toward the development of cogeneration applications from nuclear power. In tertiary oil recovery, the biggest cost factor is energy for steam, electricity, and hydrogen production, which is currently provided by natural gas. Its replacement with nuclear energy was found in a study to be competitive where HTSE as the hydrogen production method and process steam generation are considered a good option for coupling to a synthetic crude oil plant.

4.5.3 China

The demand for energy in China is rising rapidly due to its large population and its strong industrial and economic growth in recent years. Currently, China ranks first as energy consumer with the United States as second largest consumer. Coal makes up the bulk of China's primary energy consumption (66% in 2012) and will remain the dominant energy source in the next decades. Other energies consumed are oil (22%) and hydropower (12%) (USA EIA, May 2015). Due to large greenhouse gas emission, the present energy policy calls for greater energy conservation measures and a move away from coal toward cleaner energy sources including oil, natural gas, renewable energy, nuclear power, and hydroelectric resources. China has an ambitious nuclear power program for the next decades. China has 38 nuclear reactors operating with a capacity of 34.5 GW and 18 under construction with a capacity of 21 GW (March 2018). The development of HTGRs is an important part of the national nuclear program. Hydrogen activities have been enhanced in the recent years, with the main objective of improving air quality in major cities. The HTGRs are envisioned as supplement PWRs for electricity generation with more flexibility and, on the other hand, to provide process steam for heavy oil recovery or the petrochemical industry and to serve as a process heat source for coal gasification and liquefaction as well as hydrogen production.

The Chinese HTR-10 with a pebble-bed core producing a power of 10 MW(th) is one of the two HTGRs currently in operation. A follow-on demonstration plant HTR-PM will consist of two nuclear steam supply system modules, each comprising a single zone 250 MW pebble-bed modular reactor (PBMR) and a steam generator. R&D on hydrogen production through water splitting using HTGR as a process heat source was initiated in 2005 as one component of China's HTR-PM demonstration project. Both the SI thermochemical cycle and high-temperature steam electrolysis have been selected as potential processes for nuclear hydrogen production. The

Institute of Nuclear and New Energy Technology has developed SI cycle and HTSE systems for hydrogen production. The construction and testing of the SI facility were achieved at the end of 2013. The closed-cycle experiment of the SI facility was successfully carried out with the rate of 60 NL/h of hydrogen production. The desired planned outcome of the R&D of the Chinese nuclear hydrogen program is to attain a successful coupling of the hydrogen facility to the test reactor, HTR-10, which is planned to be achieved by 2020.

4.5.4 European Union

The European Union comprises highly industrialized countries with extended urban agglomerations and therefore needs to rely on a secure and economically competitive supply of energy. Primary energy by fuel share is 19% coal (down from 28% in 1990), 35% oil, 25% natural gas, 14% nuclear, and 8% renewables. The production of oil and natural gas in the European Union has been decreasing for a few years. In 2007, principal energy and climate policy targets for the European Union were redefined by the European Council to reduce greenhouse gas emission by 20% and increase the share of renewable and energy efficiencies each by 20%. The nuclear hydrogen project Reactor for Process Heat, Hydrogen, and Electricity Generation Integrated Project (RAPHAEL IP) carried out during 2005–10 consisted of 33 partners from 10 countries, with the main objectives a study of advanced gas-cooled reactor technologies needed for industrial reference designs. Hydrogen production technologies are strongly focusing on CO_2-neutral or CO_2-free methods as represented by, for example, biomass conversion or water-splitting processes or reforming of fossil fuels plus CO_2 sequestration.

Another collaborative project high-temperature thermochemical cycles (HYTHEC) (2004–08) running over almost 4 years evaluated the potential of thermochemical processes, focusing on the SI cycle to be compared with the Westinghouse hybrid (HyS) cycle. Nuclear and solar were considered as the primary energy sources, with a maximum temperature of the process limited to 950°C. The HYTHEC activities comprised, apart from the coupling to high-temperature heat sources, the modeling, chemical analysis, process flow sheeting, and experimental studies of the H_2 and O_2 production steps. The project also included industrial scale-up studies for the investigation of the feasibility of the main components, the safety aspects, and costs. The follow-up project HycycleS (2008–11) focused on providing detailed solutions for the design of specific key components for sulfur-based thermochemical cycles for hydrogen production. The key components necessary for the high-temperature part of those processes, the thermal decomposition of H_2SO_4, are a compact heat exchanger for SO_3 decomposition for operation by solar and nuclear heat, a receiver-reactor for solar H_2SO_4 decomposition, and membranes as product separator and as promoter of the SO_3 decomposition. Silicon carbide was been identified as the preferred construction material. Its stability is tested at high temperature and in a highly corrosive atmosphere. Another focus is catalyst materials for the reduction of SO_3. Requirement specifications were set up as basis for design and sizing of the intended prototypes. Rigs for corrosion tests, catalyst tests, and selectivity of separation membranes have been designed, built, and tested.

4.5.5 France

The breakdown of primary energy is 42% nuclear energy, 33% oil, 15% natural gas, 6% renewables, and 4% coal. The electricity sector in France is dominated by nuclear power, which accounted for 72.3% of total production in 2016, while renewables and fossil fuels accounted for 17.8% and 8.6%, respectively, France operates 58 nuclear power stations, the second largest fleet in the world after the United States, and has total capacity of 63.2 GW. One Gen III reactor European pressurized reactor (ERP) is currently under construction. Since nuclear energy is not always fully used, interest is growing in using excess nuclear electricity, apart from export, for hydrogen production to regulate the electricity production. France participated in different European programs on HTGR and hydrogen production.

The French AREVA NP company is developing the concept of the ANTARES reactor, a full-size HTGR design, and is a partner in the US/Russian GT-MHR project. The reference design is based on the GT-MHR of General Atomics of a 600 MW(th) reactor with prismatic block fuel and a core outlet temperature of up to 850°C for the advanced HTGR version or ~1000°C for a VHTR version. It uses an indirect cycle, possibly with a helium-nitrogen mix in the secondary system. The AREVA company has also been developing an HTGR design within the USDOE's Next Generation Nuclear Plant (NGNP) program. A comprehensive evaluation program for the ANTARES concept was completed, with a VHTR version for electricity generation and another one for process heat applications. Atomic Energy Commission (CEA) in France has been pursuing a major HTSE development program since 2004. Both planar and 3-D cells have been used for tests under typical HTSE conditions where issues such as chromium evaporation, which leads to cell pollution and thus performance loss, and thermal cycling are studied.

4.5.6 India

In recent years, energy consumption in India has been increasing at a rapid pace due to population growth and economic development. India has the world's third largest coal reserves, but the existing demand exceeds the supply. Coal accounts for 44% (as of 2013) of India's total energy consumption, followed by renewables (biomass and hydroelectric power) (26%), oil (23%), and natural gas (6%). Although nuclear power comprises only 1% of total energy consumption, it is expected to increase in the future. India's total energy needs, 30%, are met through imports. About 75% of the coal in the country is consumed in the power sector. The poor quality of Indian coal, coupled with a lack of infrastructure to clean it, poses a major environmental threat. Oil consumption accounts for roughly a third of India's energy use. India's electricity consumption totaled 1150 TWh in 2013 with 68% covered by coal, 8% by gas plants, 16% by hydropower, and 2% by nuclear. The country has 19 nuclear power plants in operation, generating 4.2 GW, and six more reactors under construction and expected to generate an additional 3160 MW. India's nuclear technology road map is seen to be strongly based on the closed fuel cycle and on the utilization of thorium, of which India has nearly one-third of the world's reserves. India has also started to work on a National Hydrogen Energy Road Map with the main targeted sectors of transportation and power supply and the longer-term goal of replacing fossil fuels with hydrogen. A government laboratory Bhabha Atomic Research Center (BARC) has started a research program on nuclear hydrogen production through water splitting based on the SI thermochemical process. BARC has successfully demonstrated SI process in closed loop operation in glass/quartz material in the laboratory. Theoretical and experimental studies are being conducted, currently concentrating on the Bunsen and the HI decomposition sections of the SI cycle.

4.5.7 Japan

Japan as one of the world's leading industrial countries has virtually no domestic oil or natural gas reserves and is the second largest net importer of crude oil and largest net importer of liquefied natural gas in the world. Total electricity production in Japan amounted to 1018 TWh in 2016, with the largest share of 41% from natural gas, 34% from coal, 7% from oil, 8% from hydro, and 8% from other renewables. Though Japan has 54 nuclear reactors with total generation capacity of 48,847 MW, all of them except for a few were shut down following 2011 March Fukushima accident and safety concerns. It is expected that these reactors will eventually restart after the safety concerns are addressed. Japan future nuclear energy plans include advanced fuel cycle and next-generation plants such as the "super-safe, small and simple" (4S) nuclear battery system, which is a sodium-cooled reactor with a steam cycle that is capable of three decades of continuous operation without refueling, or the GTHTR300, a helium-cooled VHTR for future large-scale hydrogen production. Japan is a strong player in the development of a hydrogen economy due to government support, and besides the United States, it is the leading country in fuel cell research, with large investments by both the government and the car industries.

The present H_2 consumption in Japan is about 162 million Nm^3 per year, mainly produced via SMR, and the demand for hydrogen in the future projected to rise to $54.4 \times 10^9 Nm^3$ in 2030 primarily from stationary fuel cell energy systems. Since 2005, through Japan Hydrogen and Fuel Cell (JHFC) demonstration project PEM, fuel cell vehicle (FCV) development activity has resulted in about 2000 Honda and Toyota FCV on the roads (2018). Beside fuel cells for mobile, stationary applications are being investigated. Various demonstration projects have been starting, particularly as part of complete energy systems based on renewable primary energy. With the construction and operation of the 30 MW(th) high-temperature engineering test reactor or HTTR (which achieved first criticality in 1998), the Japan Atomic Energy Agency (JAEA), formerly JAERI, has laid the basis for utilization of nuclear process heat for hydrogen production. The reactor allows a coolant outlet temperature of 950°C outside the reactor vessel to provide process heat at 905°C through the intermediate heat exchanger, which was demonstrated in 2004 for the first time in the world. Over several years, steam reforming of methane was considered the top candidate method to be connected to the HTTR. The steam reforming process was successfully demonstrated in an out-of-pile facility under nuclear conditions. In recent years, JAEA has undertaken extensive R&D on the thermochemical cycles based on the HyS and SI processes for H_2 production. It is most advanced in the study of the SI cycle, with the successful operation of a bench-scale facility having achieved a hydrogen production rate of 30 NL/h in continuous closed-cycle operation over 1 week. The next step, which started in 2005, is the design and construction of a pilot plant with a production rate of

30 Nm^3/h of H_2 under the simulated conditions of a nuclear reactor. The SI process will be designed as a "nonnuclear" grade plant; however, backup nuclear-graded equipment is available for cases of SI abnormal conditions. Based on the experience with the construction and operation of the HTTR, JAEA has developed the conceptual design of a commercial-scale HTGR hydrogen cogeneration system, GTHTR300C, which employs a helium-cooled, graphite-moderated prismatic core with 600 MW of thermal power and a helium outlet temperature of 850°C, and a gas turbine electricity generation system. The GTHTR300 employs fully passive reactor safety, high fuel burnup, a conventional steel reactor pressure vessel, nonintercooled direct Brayton cycle power conversion, a horizontal single-shaft gas turbine and electric generator, and a modular system arrangement. The reactor coolant temperature is limited to 850°C to avoid turbine blade cooling and to allow the use of conventional turbine blade materials. The elimination of an intercooling system reduces the power generation efficiency by 2%.

Toshiba and Central Research Institute of Electric Power Industry (CRIEPI) have developed the concept of the "4S reactor," a sodium-cooled pool-type reactor with 30 or 135 MW(th) power output, passive safety features, and underground construction as an additional safety measure. JAEA is also studying the concept of a sodium-cooled reactor to produce hydrogen in the so-called thermochemical and electrolytic process or JNC process. Since the maximum temperature remains below approximately 500°C and because electrolysis is applied, a fast reactor can be employed. JAEA has tested both planar and tubular versions of high-temperature electrolysis cells for hydrogen production.

4.5.8 Republic of Korea

The Republic of Korea lacks domestic energy resources and currently has to import 98% of its primary energy demand. As of 2016, the primary energy consumption share is 41% petroleum, 31% coal, 14% nuclear, 13% natural gas, and 1% renewables. Country's 31% of the electricity was generated by nuclear, 39% by coal, 19% by natural gas, 5% by petroleum, and 1% by hydropower. The Republic of Korea is a small country with a high population density where the use of low-density renewable energies is limited and not a practicable solution. A total of 20 PWR and 4 CANDU reactors provide about one-third of South Korea's electricity from 23 GWe of plant. The Republic of Korea has very active nuclear research program and is developing advanced reactors including the Korea Atomic Energy Research

Institute (KAERI) small system-integrated modular advanced reactor (SMART), a 330 MW(th) pressurized water reactor with integral steam generators and advanced safety features and designed for generating electricity (up to 100 MW(e)) and/or for thermal applications such as seawater desalination. Other advanced reactor concepts under development are a liquid metal fast/transmutation reactor and a high-temperature hydrogen generation design. A nuclear hydrogen key technology development project is underway since 2006 to develop and demonstrate the technologies required for the future nuclear H_2 by KAERI for the nuclear part and the Korea Institute of Energy Research (KIER) and the Korea Institute of Science and Technology (KIST) for the hydrogen part. The key technologies for nuclear hydrogen production investigated include VHTR design technologies, VHTR essential technologies, interfacial technologies, and coated fuel technologies. The focus of R&D on SI cycle includes the SO_3 decomposition and HI decomposition using electro dialysis.

4.5.9 Russian Federation

The Russian Federation is rich in natural energy resources. It has the largest known natural gas reserves of any country on Earth, representing 32% of the world's proved reserves. It is the world's fourth largest electricity producer after the United States, China, and Japan. In the Russian Federation, about 40% of electric power and 85% of heat supply, mainly in cogeneration, are covered by regional power industries with power plant units of ~300 MW(th). As of May 2018, Russian Federation has 35 nuclear units and a total installed capacity of 26.96 GW(e) in operation. Due to the vast amount of domestic natural gas resources and the anticipated need for large amounts of hydrogen in the future, steam reforming of methane will remain the dominant hydrogen production method in the first stage before being replaced with water-splitting production processes. Since there is the potential for considerable savings of natural gas resources when utilizing nuclear energy as a heat source for the steam reforming process, the Russian Federation has developed an increased interest in utilizing modular type HTGRs for hydrogen production. The BN-600 commercial fast reactor has the primary circuit inlet sodium temperature at 550°C. The BN-series reactors are intended for electricity generation that has been studied for other technological and commercial applications, including, potentially, commercial production of hydrogen. The reactor project MHR-T has been designed as a nuclear energy complex with a thermal power output of 600 MW(th). The complex includes a chemical-technological sector for hydrogen production plus necessary infrastructure. In the case of steam-methane reforming as the short-term option, nuclear heat is considered to be directly transferred from the primary coolant to the chemical process, while for high-temperature steam electrolysis, both two- and three-circuit plant configurations are being considered.

4.5.10 South Africa

South Africa has large coal deposits that support most of its large-scale mining and primary mineral industries. In 2014, the total primary energy consumption of the country was 181 Mtoe. The primary energy consumption mix was as follows: coal (70%), oil (23%), natural gas (3%), nuclear (3%), and renewable energies (1%). The total power generation accounted for approximately 250 TWh in 2014. Eskom Holdings Ltd., the state-owned power utility supplies about 95% of South Africa's electricity and is planning to increase the current generation capacity of 40–80 GW by 2025. The largest portion of Eskom's generation mix is coal-fired base load power. Since 1984, the national utility ESKOM has been successfully operating the Koeberg nuclear power station consisting of two 900 MW(e) PWR. The main drivers for a larger-scale national nuclear hydrogen program in South Africa are the significant domestic resources of uranium (10% of the global reserves) and the broad nuclear expertise (including HTGRs); domestic resources of platinum group metals (75% of the global reserves), required as catalyst material for fuel cells; and those of coal, providing the link to the country's highly developed coal to liquid technologies. Pebble Bed Modular Reactor (Pty) Lt (PBMR) was established in 1994, which has developed reactor concept for steam-methane reforming under the heat delivery conditions provided by an HTGR. Hot secondary helium enters the process plant at 900°C and about 9 MPa and exits at about 660°C. Temperatures are somewhat below optimum reformer performance, but still, it is believed that an HTGR can economically replace a natural gas heat source under certain conditions of high (and volatile) gas prices and the need for major reductions in CO_2 emissions. Other nuclear hydrogen production methods considered by PBMR were high-temperature steam electrolysis and the thermochemical cycles, with a preference for the Westinghouse HyS hybrid cycle.

4.5.11 United States of America

The United States consumes more energy than any other country in the world at 2450 Mteo in 2016. Fig. 4.21 shows the share of primary energy

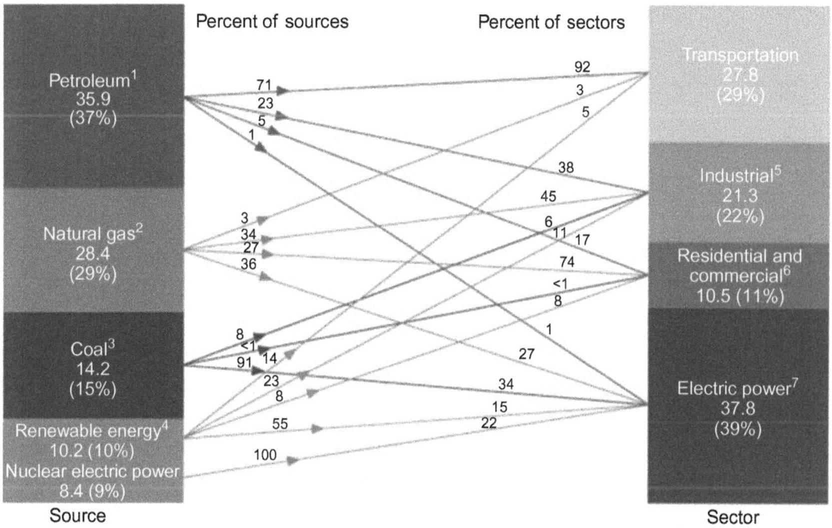

Fig. 4.21 US primary energy consumption by source and sector where it is used (U.S. Energy Information Administration, Annual Energy Outlook 2016, Department of Energy Report, DOE/EIA-0383, August 2016).

consumption and its use in the sector. Energy consumption exceeds domestic supply, and currently, the difference of the energy needs, about 14%, is imported, mainly petroleum in 2016. Coal is the most abundantly available energy resource in the United States. However, recently natural gas production has increased. The electricity generation in 2017 was about 4000 GWh. Fossil fuels are the largest sources of energy for electricity generation with natural gas producing about 32%, coal about 30%, and petroleum <1% of electricity in 2017. Nuclear energy and renewable energies including hydro produced, respectively, 20% and 17% of US electricity in 2017. The United States (March 2018) has 99 nuclear reactors at 61 operating nuclear power plants located in 30 states. Thirty-five of the plants have two or more reactors. Nuclear power has supplied about one-fifth of annual US electricity since 1990.

Total GHG emissions in 2016 was 6511 MMT of CO_2 equivalent, majority of which is contributed from transportation (28%), electricity production (28%) and industry (22%) sectors, and remaining 11% is contributed from commercial and residential sector, and 9% is from agriculture sector. The annual production of hydrogen was 13.79 million ton in 2015. Table 4.8 shows the hydrogen demand by industry in 2015.

Of all hydrogen, 95% is produced by SMR, consuming 5% of the natural gas used in the United States. Demand for hydrogen in the United States has

Table 4.8 US Hydrogen Demand by Industry

Industry H_2 Demand	MMT
Oil refining	10.542
Ammonia	2.077
Methanol	0.598
Other chemicals	0.447
Metals	0.056
Food	0.036
Electronics	0.032
Glass	0.002
Total	13.790

been increasing and is expected to grow to more than 30 million ton per year by 2030. Currently, the largest sources of hydrogen production capacity in the United States are associated with the nation's 145 operating oil refineries. Three major potential hydrogen markets seen in the United States are transportation, electric power, and production industries. R&D activities have been or are being supported by the Hydrogen Fuel Initiative of 2003; the Hydrogen, Fuel Cells, and Infrastructure Development Program; the Multiyear Research, Development, and Demonstration Plan of 2005; and the Roadmap for Manufacturing R&D on the Hydrogen Economy of 2005. In accordance with the Energy Policy Act passed in 2005, the NGNP Project consists of the research, development, design, construction, and operation of a prototype plant that (1) includes a nuclear reactor based on the research and development activities supported by the Generation IV Nuclear Energy Systems Initiative and (2) shall be used to generate electricity, to produce hydrogen, or to cogenerate electricity and hydrogen. The NGNP Project supports both the national need to develop safe, clean, and economical nuclear energy and the national hydrogen fuel initiative, which has the goal of establishing greenhouse-gas-free technologies for the production of hydrogen.

The DOE has selected the helium-cooled high-temperature gas-cooled reactor (HTGR) as the concept to be used for the NGNP because it is the most advanced Gen IV concept with the capability to provide process heat at sufficiently high temperatures for the production of hydrogen with high thermal efficiency. The NGNP Project objectives that support the NGNP mission and DOE's vision are as follows: (i) Develop and implement the technologies important to achieving the functional performance and design requirements determined through close collaboration with commercial industry end users. (ii) Demonstrate the basis for commercialization of the nuclear system, the hydrogen production facility, and the power conversion concept. An essential part of the prototype operations will be demonstrating that the requisite reliability and capacity factor can be achieved over an extended period of operation. (iii) Establish the basis for licensing the commercial version of the NGNP by the US Nuclear Regulatory Commission (NRC). This will be achieved in major part through licensing of the prototype by NRC and by initiating the process for certification of the nuclear system design. (iv) Foster rebuilding of the US nuclear industrial infrastructure and contributing to making the US industry self-sufficient for its nuclear energy production needs. In accordance with the Energy Policy Act stipulations, the NGNP preconceptual design work was focused on concepts for simultaneous electricity and hydrogen generation. US participation in GIF is on sodium-cooled fast reactors and very-high-temperature reactors.

The goal of the DOE that supported Nuclear Energy Research Initiative (NERI) was to demonstrate the commercial-scale production of hydrogen using nuclear energy. The modular helium reactor (MHR) has been suggested as the Gen IV reference concept for nuclear hydrogen generation on the basis of either the SI thermochemical cycle or HTSE. As part of the national hydrogen research program, the USDOE created the Nuclear Hydrogen Initiative (NHI) with the objective being to advance nuclear energy for the support of a future hydrogen economy. Between 2003 and 2008, the USDOE promoted nuclear hydrogen program in the United States that concentrated on the following:

- Hybrid sulfur thermochemical cycle development at Savannah River National Laboratory (SRNL).
- High-temperature electrolysis development at the Idaho National Laboratory (INL).
- SI process development at General Atomics.

As a part of the Energy Policy Act of 2005, the United States has been designing a Next Generation Nuclear Plant (NGNP). The NGNP Project's objective is to develop and demonstrate a first of a kind nuclear system with the capability to generate electric power and produce process heat for hydrogen production and other applications. The nuclear plant is based on a 400–600 MW(th) full-scale prototype gas-cooled reactor using an indirect cycle with intermediate circuit to provide electricity and process heat at 950°C. From the total thermal power, a part is planned to be consumed for hydrogen production using either the SI process or the HyS cycle or high-temperature steam electrolysis.

REFERENCES

[1] W. Doenitz, et al., Electrochemical high temperature technology for hydrogen production or direct electricity generation, Int. J. Hydrogen Energy 13 (1988) 283.
[2] W. Doenitz, et al., in: Recent advances in the development of high-temperature electrolysis technology in Germany, Proc. 7th World Hydrogen Energy Conf., Moscow, 1988, p. 65.
[3] J.E. Funk, Thermochemical hydrogen production: past and present, Int. J. Hydrog. Energy 26 (2001) 185–190.
[4] IAEA, Hydrogen as an energy carrier and its production by nuclear power, IAEA-TECDOC-1085, International Atomic Energy Agency, Vienna, 1999.
[5] IAEA, Hydrogen Production Using Nuclear Energy, IAEA Nuclear Energy Series Publications, No. NP-T-4.2, International Atomic Energy Agency, Vienna, 2013.
[6] US DOE, Nuclear Hydrogen R&D Plan, Department of Energy, Office of Nuclear Energy, Science and Technology, 2004.
[7] N.R. Brown, Study of Hydrogen Generation Plant Coupled to High Temperature Gas Cooled Reactor, PhD thesis, Purdue University, 2011.
[8] A. Botterud, B. Yildiz, G. Conzelmann, M.C. Petri, The market viability of nuclear hydrogen technologies, Argonne National Laboratory, 2007. ANL-07/13.
[9] C. Canavesio, H.E. Nassini, A.E. Bohe, Evaluation of an iron-chlorine thermochemical cycle for hydrogen production, Int. J. Hydrogen Energy 40 (28) (2015) 8620–8632.

FURTHER READING

[10] L.C. Brown, G.E. Besenbruch, J.E. Funk, A.C. Marshall, P.S. Pickard, S. K. Schowalter, High efficiency generation of hydrogen fuels using nuclear power, DE-FG03-99-SF21888; 2003, Final technical report for the period August 1, 1999 through September 30, 2002, General Atomics Corp., 2002. Report GA-A2428.
[11] N.R. Brown, S. Oh, S.T. Revankar, C. Kane, S. Rodriguez, R. Cole Jr., et al., Analysis model for sulfur-iodine and hybrid sulfur thermochemical cycles, J. Nucl. Technol. 166 (2009) 4355–4361.
[12] N.R. Brown, V. Seker, S.T. Revankar, T.J. Downar, An endothermic chemical process facility coupled to a high temperature reactor. Part I: Proposed accident scenarios within the chemical plant, Nucl. Eng. Des. 246 (2012) 256–265.
[13] R. Elder, R. Allen, Nuclear heat for hydrogen production: coupling a very high/high temperature reactor to a hydrogen production plant, Prog. Nucl. Energy 51 (2009) 500–525.
[14] R.S. El-Emam, I. Khamis, Advances in nuclear hydrogen production: results from an IAEA international collaborative research project, Int. J. Hydrog. Energy (2018), https://doi.org/10.1016/j.ijhydene.2018.04.012.
[15] Govt of India, Draft report on hydrogen production in India, Sub-Committee on Research, Development & Demonstration for Hydrogen and Fuel Cells, Ministry of New and Renewable Energy, New Delhi, 2016.
[16] JAERI, n.d. High-temp Engineering test reactor (HTTR) used for R&D on diversified, application of nuclear energy http://www.jaeri.go.jp/english/ff/ff45/tech01.html.
[17] I. Khamis, U.D. Malshe, HEEP: a new tool for the economic evaluation of hydrogen economy, Int. J. Hydrogen Energy 35 (2010) 8398–8406.
[18] G.F. Naterer, et al., Nuclear energy and its role in hydrogen production. Chapter 2, in: Hydrogen Production From Nuclear Energy, Springer-Verlag, London, 2013, pp. 21–64. https://doi.org/10.1007/978-1-4471-4938-5_2.
[19] P. Nikolaidis, A. Poullikkas, A comparative overview of hydrogen production processes, Renew. Sust. Energ. Rev. 67 (2017) 597–611.

[20] J.H. Norman, G.E. Besenbruch, L.C. Brown, D.R. O'Keefe, C.L. Allen, Thermochemical water-splitting cycle, bench scale investigations, and process engineering. Final report for the period February 1977 through December 1981, GA-A16713, General Atomics Corp, La Jolla, CA, 1982.
[21] J.E. O'Brien, C.M. Stoots, J.S. Herring, G.L. Hawkes, Hydrogen production from nuclear energy via high temperature electrolysis, 1st Energy Center Hydrogen Initiative Symposium INL/CON-06-01375, April 2006, 2016.
[22] OECD, in: Nuclear Production of Hydrogen, Nuclear Energy Agency Organisation for Economic Co-Operation and Development, First Information Exchange Meeting, Paris, France, 2–3 October, 2000.
[23] S.T. Revankar, Transient analysis of coupled high temperature nuclear reactor to a thermochemical hydrogen plant, Int. J. Hydrog. Energy 38 (2013) 6174–6181.
[24] S.T. Revankar, N.R. Brown, C. Kane, S. Oh, Development of efficient flowsheet and transient modeling for nuclear heat coupled sulfur iodine cycle for hydrogen production, DEFC0706ID14749, Project No. 06-060; 2010. Final technical report for period March 13, 2007–March 12, 2010, Purdue University, 2010. PU-NE 10-23.
[25] H. Sato, H. Ohashi, S. Nakagawa, Y. Tachibana, K. Kunitomi, Safety design consideration for HTGR coupling with hydrogen production plant, Prog. Nucl. Energy 82 (2015) 46–52.
[26] S. Suppiah, Hydrogen Production and Utilization A Canadian Perspective – Canadian Nuclear Laboratories, Chalk River, Canada, Pathways to Decarbonization: An International Workshop to Explore Synergies Between Nuclear and Renewable Energy Sources National Renewable Energy Laboratory, Golden, Colorado June 9–10, 2016.
[27] T. Trygve Riis, E.F. Hagen, P.J.S. Vie, O. Ulleberg, Hydrogen Production and Storage: R&D Priorities and Gaps, Organisation for Economic Co-Operation and Development, International Energy Agency, Paris, France, 2006.
[28] US DOE, Report of the Hydrogen Production Expert Panel: A Subcommittee of the Hydrogen & Fuel Cell Technical Advisory Committee, United States Department of Energy, Washington, DC, 2013.
[29] US NRC, Nuclear Power Plant Licensing Process, http://www.nrc.gov/reactors/new-reactor-licensing.html., May 14, 2018.
[30] T.S. Veras, T.S. Mozer, D.C.R.M. Santos, A.S. Cesar, Hydrogen: trends, production and characterization of the main process worldwide: review Article, Int. J. Hydrogen Energy 42 (2017) 2018–2033.
[31] X.L. Yan, R. Hino, Nuclear Hydrogen Production Handbook, CRC Press, Taylor & Francis Group, Boca Raton, FL, 2011.
[32] B. Yildiz, M.S. Kazimi, Efficiency of hydrogen production systems using alternative nuclear energy technologies, Int. J. Hydrog. Energy 31 (2006) 77–92.
[33] P. Zhang, L. Wang, S. Chen, J. Xu, Progress of nuclear hydrogen production through the iodine–sulfur process in China, Renew. Sust. Energ. Rev. 81 (2018) 1802–1812.

CHAPTER FIVE

Selecting Favorable Energy Storage Technologies for Nuclear Power

Samuel C. Johnson*, F. Todd Davidson*, Joshua D. Rhodes*, Justin L. Coleman†, Shannon M. Bragg-Sitton‡, Eric J. Dufek§, Michael E. Webber*

*The University of Texas at Austin, Austin, TX, United States
†Fusion Hydrogen and Measurement Science Department, Idaho National Laboratory, Idaho falls, ID, United States
‡Systems Integration Department, Nuclear Systems Design and Analysis, Idaho National Laboratory, Idaho falls, ID, United States
§Energy Storage and Advanced Vehicles Department, Idaho National Laboratory, Idaho falls, ID, United States

5.1 INTRODUCTION

In recent years, growth in electricity generation from variable renewable energy sources and inexpensive natural gas has been significant [1]. Market deregulation has led to an environment in which nuclear power plants that have traditionally operated at close to full capacity have been called upon to operate more flexibly and compete directly with newer plants [2]. Consequently, concerns have been raised that nuclear power plants will need to adapt to the new paradigm or risk being shuttered due to disadvantageous economics [3]. One proposed solution to this challenge is integrating energy storage, which could increase available revenue by enabling more flexible operation while reducing operational costs by stabilizing power output, which lowers maintenance expenses and improves safety by avoiding unnecessary thermal stresses from cycling [4,5]. However, the appropriate energy storage technology for enhancing the flexibility of a nuclear power plant is not immediately obvious and will depend on a variety of site-, plant- and market-specific factors. The compatibility of an energy storage solution can be measured across many metrics, including environmental impact, technical maturity, and cost.

Nuclear power plants (NPPs) provide the US electricity grid with a substantial fraction of total generation (approximately 20%) [6] and an even larger fraction of its low-carbon power (almost 60%) [7]. Traditional

Storage and Hybridization of Nuclear Energy
https://doi.org/10.1016/B978-0-12-813975-2.00005-3

operation of NPPs provides the grid with stable electricity generation throughout the day while producing less greenhouse gas emissions over the life cycle of the plant than other electricity-generating units. On average, nuclear reactors produce 66 gCO_2e/kWh in life-cycle emissions, mostly from plant construction and fuel preparation (e.g., mining, enrichment, and fabrication), which is almost an order of magnitude less than natural gas combined-cycle turbines [8]. Despite these benefits, ambiguous national policies affecting the nuclear power industry, low marginal prices from natural gas and renewable sources, and large, multidecade capital investments required to replace or retrofit the nuclear fleet have resulted in a difficult economic climate for NPPs [9, 10]. Hence, electricity generation from nuclear reactors has not increased since the 1990s, while electricity generation from natural gas, wind, and solar has grown considerably, as shown in Fig. 5.1 [11].

NPPs have been built globally, but new NPPs in the United States were not constructed for 2 decades during the period of 1996–2016. There has been renewed investment in nuclear power in recent years: The first new reactor in the United States since 1996, Watts Bar 2, went online in October 2016, and two additional reactors are projected to begin operation by 2021.

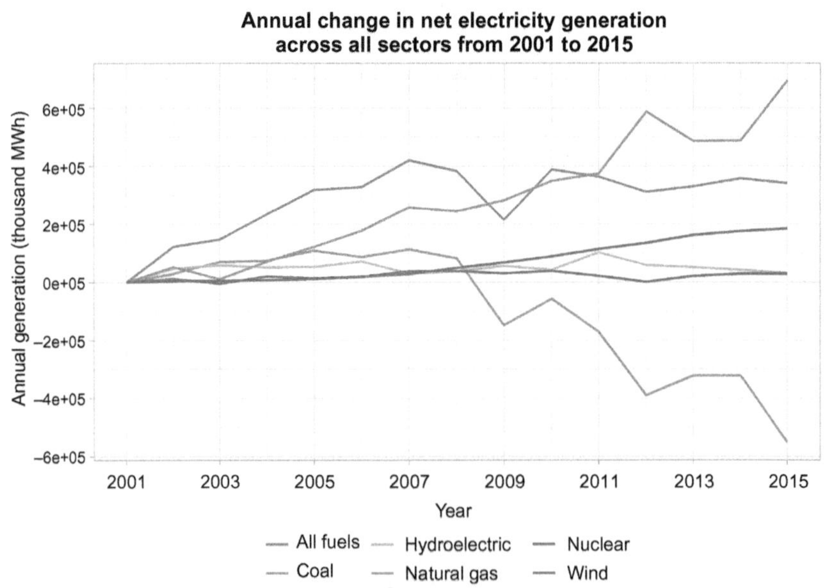

Fig. 5.1 Change in net electricity generation across all sectors from 2001 to 15 shows decreases in coal, increases in wind and natural gas, and stable output from hydro- and nuclear power plants [11].

The development of small modular reactors (SMRs) that might have less investment risk, introduction of direct policy support for NPPs, or a more aggressive policy stance toward reducing carbon emissions could foster a more favorable investment climate for NPPs [12]. The IAEA projects that nuclear power will continue to grow worldwide, with estimates of up to 42% growth by 2030 as electricity demand in Asia rises and nations attempt to meet their commitments under 2015's Paris Agreements. Though, expectations since the Fukushima disaster in 2011 have been more conservative due to price competition and stricter safety standards [13].

Continued reduction of carbon emissions from electricity generation in the United States will likely require multiple avenues of remediation [14]. While it might be possible [15], it is difficult and potentially cost-prohibitive for renewable energy sources to provide all of the carbon-free electricity generation in the United States due to resource variability and geographic dispersion [16]. Thus, it is worth investigating whether energy storage could supplement renewable electricity generation to help meet climate change mitigation goals by accommodating variability in renewable output while also improving the economic outlook for NPPs. Energy storage technologies could improve NPP performance by providing NPPs with multiple avenues for generating revenue, such as the delivery of ancillary services [17]. An integrated energy storage system could also bolster the resilience of NPPs against market trends by enhancing the flexibility of the plant, allowing operators to more easily ramp their supply of electricity to the grid to match changes in demand. Although the ramp rates of NPPs are typically constrained by economics or regulations rather than technical ability, the advantages of flexibility have been proved in the literature [18], and energy storage might provide plant operators a way to overcome these economic and regulatory barriers. Energy storage could also help reduce maintenance costs induced by thermal stresses from plant cycling [4,5]. In this chapter, several energy storage technologies are compared as potential candidates for near-term installation alongside newly constructed advanced nuclear power plants. Advanced NPPs are defined here as the Generation III+ reactors currently being built in the United States, including the Westinghouse AP1000 and GE-Hitachi ABWR reactor designs [19].

Generation III+ reactor designs, or *advanced NPPs* in this analysis, have many features that set them apart from previous generations. These reactors prioritize simpler systems that are intended to reduce capital costs and are also more fuel-efficient and safety-conscious. Other unique features of advanced NPPs include more standardized designs to simplify licensing, expectation for longer operating lives, higher availability due to a more robust design and some load-following capabilities, higher burnup, and the implementation of passive safety systems [19]. In this analysis, advanced NPPs were limited to commercial reactors >700 MWe in design, so SMRs were not considered. The Westinghouse AP1000 reactor was explored in greatest detail in this chapter due to its popularity in the United States. The AP1000 reactor was designed to ramp electricity generation by 5% of the plant's nameplate capacity per minute between 15% and 100% of the plant's maximum power output, but integrated energy storage could help further reduce operation and maintenance (O&M) costs and increase the plant's total output. The same principles for integrating energy storage with an advanced NPP apply to more traditional NPPs currently in operation [20].

Energy storage still faces many barriers to widespread adoption that need to be addressed, with different technologies in varying levels of development. Each technology's development timeline should be an important consideration when selecting compatible energy storage systems. Every storage technology considered for integration with an advanced NPP should also be assessed for its environmental impact [21]. Additionally, although energy storage technologies such as batteries are currently experiencing cost reductions, current market conditions indicate that costs still need to improve before they will be able to provide economic benefit to an advanced NPP [22]. For this reason, several thermal energy storage technologies were also considered in this chapter, since these energy storage systems can be 10–40 times less expensive than electricity storage [23]. In this chapter, energy storage characteristics were matched to the specifications of Vogtle 3 and 4, two reactors currently under construction in Waynesboro, Georgia, to identify the most favorable technologies for integration with this representative power plant. Nineteen different energy storage technologies are compared in this chapter, all of which are listed in Table 5.1.

5.2 DESCRIPTIONS OF THE CONSIDERED ENERGY STORAGE TECHNOLOGIES

The energy storage technologies discussed in this chapter are described in more detail here. Additional parameters describing the operation of these systems are recorded in Appendix A.

Table 5.1 The Storage Technologies Considered and Compared in This Chapter Represent a Comprehensive But Not Exhaustive List of Available Technologies

Type of Energy Storage	Energy Storage Technology
Mechanical	Pumped-storage hydropower
	Compressed-air energy storage
	Flywheels
Electric	Supercapacitors
	Superconducting magnetic energy storage
Electrochemical (conventional batteries)	Lithium-ion
	Sodium-sulfur
	Lead-acid
	Nickel-cadmium
Electrochemical (flow batteries)	Zinc-bromine
	Vanadium redox
Chemical	Hydrogen
Thermal (sensible heat)	Underground thermal energy storage
	Hot/cold water
	Solid media
Thermal (latent heat)	Thermochemicals
	Molten salts
	Liquid air
	Phase change materials

5.2.1 Mechanical Energy Storage

5.2.1.1 Pumped Storage Hydropower

Pumped-storage hydropower (PSH) is the most developed energy storage technology in the world today. The IEA estimates that PSH installations account for 99% of the energy storage capacity worldwide [24]. In the United States, the PSH fleet consists of 42 plants accounting for 21.6 GW of capacity, or 97% of the total utility-scale electricity storage in the United States at the end of 2015 [25]. The construction of new PSH facilities in the United States stalled in the mid-1980s due to environmental opposition and the changing needs of the grid, triggered by the transition to restructured electricity markets [26]. However, models built by the DOE have shown that there is potential for 35 GW of additional PSH facilities to be installed by 2050, which would more than double the current capacity in the United States [25]. PSH plants store energy by pumping water from a lower reservoir to a higher reservoir using electricity generated during off-peak periods. During peak demand periods, the water flows back down to the lower reservoir, generating electricity. PSH facilities can offer developers better ramp rates than natural gas power plants for increasing the flexibility of the grid. The environmental impacts of PSH facilities are significant, though, and specific geographic conditions must be available to make construction viable. The investment costs for PSH plants can also be prohibitive [26].

5.2.1.2 Compressed Air Energy Storage

Compressed-air energy storage (CAES) facilities have been commercially deployed, but are not nearly as widespread as PSH plants. Only two full-scale CAES systems are in operation in the world today: one in Germany and one in the United States [26]. Like PSH plants, specific geographic formations are typically required for CAES installations. A CAES system stores energy by using off-peak electricity to compress air and store it in a reservoir. Although large, steel, aboveground containers can be built to use as a reservoir for the compressed air, naturally occurring salt caverns often provide a more cost-effective alternative. The compressed air is heated, expanded, and released to a combustor in a gas turbine during peak demand periods to generate electricity. CAES plants offer quick ramp rates like PSH facilities, but the efficiency of the energy storage and conversion process is relatively low compared with other energy storage technologies. Likewise, CAES plants are slower to respond to disruptions in the grid than quick-response technologies like flywheels or batteries [27].

5.2.1.3 Flywheels

Flywheels store kinetic energy with a spinning rotor. Controls and a power conversion system are used to convert AC power delivered by the grid or an individual power plant into the rotational energy of the rotor. The energy is later released by applying resistance to the spinning rotor. Flywheels have very low energy capacities compared with PSH and CAES systems but can deliver much more power per mass. In modern flywheel systems, the spinning rotor is contained in a thick, steel vessel that protects the rotor and the motor-generator used to convert electric energy into mechanical energy and vice versa. This containment vessel also protects surrounding workers from injury in the event of a catastrophic failure. The rotor is also surrounded by a vacuum to minimize the frictional loss of energy as the rotor spins [24,26]. Flywheels offer many benefits to developers, since they are a durable, modular, and quick-responding technology. Flywheel energy storage systems are also highly efficient, and their scalability to grid-scale applications has been proved [27].

5.2.2 Electrical Energy Storage

5.2.2.1 Supercapacitors

Capacitors store energy by collecting positive and negative charge on two conductive plates opposite one another and separated by a dielectric material. An electric field forms between the two plates that can be used to quickly store and release electricity. Supercapacitors, which are also called electric double-layer capacitors, usually have an energy density hundreds of times greater than that of a conventional capacitor. Supercapacitors store energy between two high-surface-area electrodes separated by an ion-permeable membrane. An electrolyte solution is used to carry charge between the two electrodes. The large surface area of supercapacitor electrodes allows for higher energy density but has the drawback of lower power density compared with conventional capacitors [28,29]. Compared with electrochemical batteries, supercapacitors could be characterized as having high power densities and low energy densities [27]. Figs. C.1–C.5 in Appendix C provide information on the relative power and energy density performance of supercapacitors versus electrochemical batteries.

5.2.2.2 Superconducting Magnetic Energy Storage

Superconducting magnetic energy storage (SMES) systems store energy in a magnetic field. This magnetic field is generated by a DC current traveling through a superconducting coil. In a normal wire, as electric current passes through the wire, some energy is lost as heat due to electric resistance. However, in a SMES system, the wire is made from a superconducting material that has been cryogenically cooled below its critical temperature. As a result, electric current can pass through the wire with almost no resistance, allowing energy to be stored in a SMES system for a longer period of time. Common superconducting materials include mercury, vanadium, and niobium-titanium. The energy stored in an SMES system is discharged by connecting an AC power convertor to the conductive coil [30]. SMES systems are an extremely efficient storage technology, but they have very low energy densities and are still far from being economically viable [27].

5.2.3 Electrochemical Energy Storage (Conventional Batteries)

5.2.3.1 Lithium-Ion Batteries

In recent years, lithium-ion batteries have been used as the energy storage technology of choice for consumer products, electric vehicles, personal electronics, and many other applications in which the weight of the energy storage technology needs to be minimized. In the past couple of years, lithium-ion batteries have also dominated the market for stationary grid-scale energy storage applications [26]. In a lithium-ion battery cell, energy is stored by causing positively charged lithium ions to travel through a liquid electrolyte to the opposing electrode, while electrons are transferred through an external circuit. When the battery is discharging, lithium ions are transferred back to the original electrode allowing for a discharge of energy through the external circuit. Lithium has a high galvanic potential, giving lithium-ion batteries favorable energy storage characteristics. However, lithium is also highly reactive when exposed to oxygen or water and must be packaged carefully. The lifetime and costs for this technology are not as favorable as other energy storage technologies, but lithium-ion batteries offer superior energy density and specific energy characteristics when compared with other commercially available electrochemical storage options. Thus, lithium-ion batteries are still the storage technology of choice for many mobile devices [29].

5.2.3.2 Sodium-Sulfur Batteries

Proponents of sodium-sulfur (NaS) battery systems claim that this technology is the most economically feasible battery storage option available, though NaS battery systems are like other battery systems in many ways. This technology's defining characteristic is its long discharge period, which can exceed 6h. NaS batteries require careful maintenance due to their extreme operating conditions. In a NaS battery, molten sodium and sulfur act as the battery's two electrodes, with beta-alumina acting as the solid electrolyte. Sodium ions layered in aluminum oxide carry charge across the electrolyte. Therefore, the operating temperature of the battery must be kept between 300°C and 350°C. The sodium must also be prevented from coming into contact with water and combusting. The high operating temperatures coupled with the high reactivity of the component elements used in NaS batteries have led to the implementation of strict safety measures [26]. Despite the safety challenges, NaS batteries offer superior energy densities and show promise for use in applications that require short and powerful bursts of energy [30].

5.2.3.3 Lead-Acid Batteries

Lead-acid batteries were the first rechargeable electrochemical battery storage available. This storage technology was first developed in the mid-1800s and was soon adopted for commercial applications. In a lead-acid battery, the cathode is made of lead-dioxide, and the anode is made of metallic lead. The two electrodes are separated by an electrolyte of sulfuric acid. As the battery charges, the sulfuric acid reacts with the lead in the anode and cathode to produce lead sulfate. A reverse process occurs when the battery is discharging. The production and decomposition of this chemical produce short and powerful bursts of energy, enough to start a car, boat, or plane. However, the gradual crystallization and buildup of lead sulfate in the battery's core severely reduce the cycle life of these batteries. As a result, they are not an ideal technology for several energy management services [26,28–30]. Due to their low energy density, this technology also has a larger footprint than other batteries [27].

5.2.3.4 Nickel-Cadmium Batteries

Nickel-cadmium (NiCd) batteries are direct competitors with lead-acid batteries since these batteries offer similar technical characteristics but with superior cycling abilities and energy density. In a NiCd battery, nickel oxide hydroxide is used to make the cathode, and the anode is made from metallic cadmium. An aqueous alkali solution is used as the electrolyte between the two electrodes. NiCd batteries are currently widely used for portable electronics applications, like lead-acid and lithium-ion batteries. Despite their superior cycling characteristics and energy density, NiCd batteries have their drawbacks. The batteries are constructed from highly toxic materials and suffer from the "memory effect," which requires that the battery be fully recharged even after a partial discharge [28–30].

5.2.4 Electrochemical Energy Storage (Flow Batteries)

5.2.4.1 Zinc-Bromine Flow Batteries

Zinc-bromine (ZnBr) flow batteries can be categorized as hybrid flow batteries, which means that some of the energy is stored in the electrolyte and some of the energy is stored on the anode by plating it with zinc metal during charging. In a ZnBr battery, two aqueous electrolytes act as the electrodes of the battery and store charge. The electrolyte solutions contain the reactive components, zinc and bromine, and as these solutions flow through the battery's cells, reversible electrochemical reactions occur, and energy is either charged to the battery or discharged. When the battery is charging, elemental zinc attaches to the carbon-plastic electrodes connecting each cell in the battery to form the anode, and bromine forms at the cathode. Carbon plastic is used for the electrodes because of the highly corrosive nature of bromine. A selective membrane is included in the battery's design to separate the electrolytes while still allowing ion transfer to maintain charge neutrality [26,29,30]. Flow batteries have many advantages including long lifetimes, modularity, and almost no energy loss throughout the technology's storage duration. However, the design for these battery systems can be very complex, which can lead to increased costs and difficulties in development [27,28]. ZnBr flow batteries also feature lower efficiencies and stricter operating conditions than most other battery storage technologies [30].

5.2.4.2 Vanadium Redox Flow Batteries

In contrast to ZnBr flow batteries, vanadium redox batteries (VRBs) only store energy within the electrolyte of the battery. VRBs are the most mature type of flow battery available. They were first developed in the 1980s and now constitute over 20 MWh of installed storage capacity worldwide. VRBs are used mostly for small- and medium-scale applications, but their utility in responding to variable generation from renewable energy resources has already been demonstrated. VRBs store energy with vanadium redox couples that are kept in two separate electrolyte tanks. As the electrolyte flows through the battery during charging, vanadium ions accept electrons at the anode and deposit electrons at the cathode. The reactions run in the reverse direction when the battery is discharging. As with ZnBr flow batteries, a proton-exchange membrane is needed to allow charge to flow, while the electrolyte solutions are kept separate. A significant advantage of VRB systems is that the two electrolyte solutions are chemically identical, which makes the operation of the battery much simpler and less expensive [26,28–30]. However, VRBs still face technical challenges, including low electrolyte stability and solubility, which can lead to decreased energy densities. The operating costs for VRBs also remain too high for the technology to be economically viable [30].

5.2.5 Chemical Energy Storage

5.2.5.1 Hydrogen Energy Storage

The production of hydrogen for energy storage is different than many of the other technologies considered in this report. First, rather than simply charging an energy storage device directly, hydrogen must be produced from an alternative resource. Hydrogen can be produced through the electrolysis of water using electricity produced by a nearby power plant or another electricity-generating unit. An electrolyzer introduces an electric current to the water to produce hydrogen and oxygen [28,30]. Two primary electrolysis technologies are currently available, alkaline electrolysis and proton-exchange-membrane (PEM) electrolysis, and both operate at relatively low temperatures (<100°C). However, high-temperature hydrogen production methods (700–900°C) are being researched and could be more compatible with nuclear power plants [31]. After the hydrogen is produced, it must be stored or used for another application. Possible postproduction uses include power-to-power, when hydrogen is stored in an underground cavern or pressurized tank to be converted to electricity later using either a fuel cell or a hydrogen-fueled gas turbine. Other postproduction uses include power-to-gas, when hydrogen is either blended with natural gas or used to create synthetic methane; power-to-fuel, when the hydrogen is used as a fuel for the transportation sector; and finally power-to-feedstock, when produced hydrogen is used for chemical and refining industries [32]. Although hydrogen production is a versatile energy storage method, offering clean and efficient electricity generation as well as scalability and a compact design, many challenges still face this technology. The primary limitations of hydrogen energy storage systems are the durability of the system components, high investment costs, and possible geographic requirements related to the hydrogen storage vessel [28,30].

5.2.6 Thermal Energy Storage (Sensible Heat)

5.2.6.1 Underground Thermal Energy Storage

Underground thermal energy storage (UTES) systems store energy by pumping heat into an underground space. There are three typical underground locations in which thermal energy is stored: boreholes, aquifers, and caverns or pits. The storage medium typically used for this method of thermal energy storage is water. Boreholes are man-made vertical heat exchangers that work to transfer heat between the energy carrier and the ground layers. Conversely, aquifers and underground caverns or pits are natural storage spaces for thermal energy. In aquifers, thermal energy is transferred to the aquifer by injecting or extracting hot or cold water from the aquifer itself. Finally, thermal energy stored in underground caverns or pits is stored in a large underground reservoir. Although, this last form of underground thermal energy storage is technically feasible, installations have been limited due to high investment costs. Additionally, although UTES systems are a convenient form of bulk thermal energy storage, their success is largely dependent on surrounding geographic conditions and a local need for district heating. UTES systems are incapable of contributing to high-temperature applications since it is impractical to store water underground above its standard boiling temperature for typical operating pressures [24,33].

5.2.6.2 Hot/Cold Water Storage

Hot and cold water storage tanks are probably the most prominent form of thermal energy storage. These energy storage systems are used primarily to shift the energy demand for the heating and cooling of residential and commercial buildings to off-peak periods to reduce costs. There are many different versions of this technology. For example, domestic water heaters can be used as a distributed form of thermal energy storage. In France, the thermal energy storage capacity available in domestic electric water heaters is responsible for reducing the country's peak energy demand in the winter by about 5%. By allowing the utilities to gain control over individual water heaters throughout the country, the peak energy demand can be reduced, and costs are returned to the consumer [24]. Steam accumulators are another form of hot water energy storage in which steam produced by a power plant is stored directly as a pressurized saturated liquid [34]. In a typical thermal energy storage system using hot or cold water, the device chills or transfers heat to the water, which is then stored in an insulated tank. The water is held at temperatures either right above the freezing temperature of water or right below the boiling temperature. Pressurized storage tanks can hold water at even higher temperatures. Even still, the storage output temperature of this technology is severely limited [24,29].

5.2.6.3 Solid Media Storage

Water has a very high heat capacity, and as a result, water has a high energy storage density. However, as a form of sensible thermal energy storage, water also has limitations. Since the boiling and freezing temperatures for water are relatively close compared to other materials, such as concrete, water can only be heated to a certain temperature without causing it to boil, and it can only be cooled so much before it begins to freeze. Freezing or boiling water can have drawbacks because water is often transported as a liquid through pipes and stored in tanks or, in the case of UTES, underground caverns and aquifers. Solid media energy storage systems offer a form of sensible thermal energy storage for high-temperature applications. Common solid materials used for thermal energy storage include concrete, bricks, and rocks. These materials are inexpensive, environmentally friendly, and easy to handle. The energy density of solid materials is generally much lower than liquid storage media though [24,33]. Energy is usually transferred to a solid storage medium by first transferring the thermal energy to some heat transfer fluid that runs alongside the solid storage medium as in a conventional heat exchanger [35]. The solid storage medium could also be electrically heated, as with firebrick thermal energy storage systems [36].

5.2.7 Thermal Energy Storage (Latent Heat)

5.2.7.1 Thermochemicals

Thermochemical storage (TCS) systems have emerged as a potential energy storage solution recently due to the technology's superior energy density and absence of energy leakage throughout the technology's storage duration. TCS systems store energy in endothermic chemical reactions, and the energy can be retrieved at any time by facilitating the reverse, exothermic reaction. The storage output temperature is dependent on the properties of the thermochemical that was used as the storage medium [24]. Typically, thermochemical energy storage refers to two main processes, thermochemical reactions and sorption processes. Thermal adsorption reactions can be used to store heat or cold in the bonding of a substance to another solid or liquid. A common sorption process used in TCS systems is the adsorption of water vapor to silica gel or zeolites. During charging, the water is desorbed from the inner surface of the adsorbent and is adsorbed again when the stored energy is discharged from the system [33]. Alternatively, heat can be stored by directing thermal energy to an endothermic chemical reaction. In this reaction, a thermochemical absorbs the energy and splits into separate substances, which can be stored until the energy is needed again. The reverse reaction occurs when the two substances are recombined and thermal energy is released through this exothermic reaction. The latent heat of the reaction for the selected thermochemical is equal to the storage capacity of the system [37]. Although the energy densities of thermochemicals are greatly superior to other energy storage technologies, thermochemicals are currently economically infeasible [27].

5.2.7.2 Molten Salts

Molten salts are a phase change material that is commonly used for thermal energy storage. Molten salts are solid at room temperature and atmospheric pressure but change to a liquid when thermal energy is transferred to the storage medium. In most molten salt energy storage systems, the molten salt is maintained as a liquid throughout the energy storage process. Molten salts are typically made up of 60% sodium nitrate and 40% potassium nitrate, and the salts melt at approximately 220°C [29]. Molten salts are often used with concentrating solar power (CSP) plants to store thermal energy for electricity generation [24]. In CSP plants, excess heat that is not used for electricity generation is diverted to the molten salt, which is then stored in an insulated tank. After sunset, this thermal energy can be used to produce steam and generate electricity when the sun is no longer providing energy to the CSP plant. This thermal energy storage capacity can also be used to smooth electricity production throughout the day and mitigate the variability associated with solar PV technologies [38]. In fact, the integration of thermal energy storage capacity can increase the capacity factor of a CSP plant from 25% to nearly 70% [29].

5.2.7.3 Liquid Air

Liquid air energy storage (LAES) technologies are gaining traction as an efficient and cost-effective energy storage method due to their large scale and long duration as well as their compatibility with existing infrastructure. LAES systems store energy using a method similar to CAES systems. Instead of storing compressed air in a large cavern, though, the volume of the gas is reduced further by refrigerating the air and liquefying it. The liquid air is then stored in an insulated, low-pressure tank aboveground, eliminating the geographic requirements associated with CAES systems. In LAES systems, natural gas is typically burned to drive the expansion process. However, the advanced adiabatic and isothermal compression methods that are being developed for CAES systems are applicable to LAES systems as well. Utilizing waste heat or cold from other processes, such as LNG terminals or landfill gas engines, could further improve the efficiency of this technology and eliminate the need for an external energy source [39].

5.2.7.4 Phase Change Materials

Although sensible thermal energy storage can be effective and relatively inexpensive, latent thermal energy storage technologies offer superior energy densities and target-oriented discharge temperatures. Molten salts and liquid air are both specific types of phase change materials (PCMs) that have developed into independent technologies due to their technical maturity. In theory, any PCM can be used for thermal energy storage, but only a few have been proved as effective. With PCMs, as thermal energy is transferred to or away from the chosen storage medium, the material changes phase. Since all of the thermal energy transferred to the material is directed to changing the material's phase, PCMs absorb and release heat isothermally. Depending on the material used, PCM thermal energy storage systems can be used for either shifting daily energy time-of-use or seasonal energy storage. However, although these materials can store 5–14 times more thermal energy per unit volume than sensible energy storage technologies, a phase change material must have very specific properties to be an effective storage medium. For example, to be used as latent heat storage medium, PCMs should have a phase-transition temperature that aligns well with the desired operating temperature, a high latent heat of fusion, and a high thermal conductivity. These materials can also be expensive and rare, which could slow the technology's progression toward maturity [33,40].

5.3 COMPREHENSIVE COMPARISON OF ENERGY STORAGE TECHNOLOGIES

Several diverse data sets regarding energy storage performance characteristics were curated as part of this work (see Appendix A). The performance data and unique energy storage characteristics were categorized into five principal bins (described further below): technical maturity, economic feasibility, environmental impact, logistic constraints, and policy and market considerations.

5.3.1 Technical Maturity

The maturity of each energy storage technology was determined using a widely accepted assessment technique originally developed by NASA and DOD that has been tailored by the DOE for application to energy-related technologies. This framework is called the *technology readiness assessment* and is useful for assigning a *technology readiness level* (TRL) to a technology on a scale of 1–9 [41]. This framework is recorded in Table 5.2.

Each storage technology considered in this analysis was assigned a TRL score between 1 and 9 that corresponds to the technology's current stage in development, based on the methodology developed by the DOE. This score allows for easy comparison of energy storage technologies. Technical maturity is important for NPPs close to beginning operation or when considering retrofitting existing NPPs, but might not matter so much for NPPs early in their development timeline.

5.3.2 Economic Feasibility

Energy storage technologies often have widely varying energy and power costs. For example, one technology might have higher power-related costs (e.g., generation assets for pumped hydropower energy storage), while another technology could have higher energy-related costs (e.g.,

Table 5.2 DOE Technology Readiness Level Framework [41]

Relative Level of Technology Deployment	Technology Readiness Level	TRL Definition	Description
System operations	TRL 9	Actual system operated over the full range of operating conditions	Actual operation of the technology in its final form, under the full range of operating conditions
System commissioning	TRL 8	Actual system completed and qualified through test and demonstration	Technology has been proved to work in its final form and under expected conditions. In almost all cases, this TRL represents the end of true system development
	TRL 7	Full-scale, similar (prototypical) system demonstrated in a relevant environment	Prototype full-scale system. Represents a major step up from TRL 6, requiring demonstration of an actual system prototype in a relevant environment
Technology demonstration	TRL 6	Engineering/ pilot-scale, similar (prototypical) system validation in a relevant environment	Representative engineering-scale model or prototype system, which is well beyond the lab scale tested for TRL 5, is tested in a relevant environment. Represents a major step up in a technology's demonstrated readiness
Technology development	TRL 5	Laboratory scale, similar system validation in relevant environment	The basic technological components are integrated so that the system configuration is similar to (matches) the final application in almost all respects
	TRL 4	Component and/ or system validation in laboratory environment	Basic technological components are integrated to establish that the pieces will work together. This is relatively "low fidelity" compared with the eventual system

Table 5.2 DOE Technology Readiness Level Framework [41]—cont'd

Relative Level of Technology Deployment	Technology Readiness Level	TRL Definition	Description
Research to prove feasibility	TRL 3	Analytic and experimental critical function and/or characteristic proof of concept	Active research and development is initiated. This includes analytic studies and laboratory sale studies to physically validate the analytic predictions of separate elements of the technology
Basic technology research/ research to prove feasibility	TRL 2	Technology concept and/or application formulated	Invention begins. Once basic principles are observed, practical applications can be invented. Applications are speculative, and there may be no proof or detailed analysis to support the assumptions
Basic technology research	TRL 1	Basic principles observed and reported	Lowest level of technology readiness. Scientific research begins to be translated into applied research and development (R&D)

supercapacitors) [22]. It can be difficult to directly compare the costs of different storage technologies for this reason. In this chapter, the full cost of installation was estimated for each energy storage option by calculating the cost of fulfilling a specific application's power and energy requirements. Since both per-kW and per-kWh capacity costs represent the full cost of investment for an energy storage system, the largest of these two *full costs* was estimated as representative of the investment cost for that technology. This allows for an objective comparison of energy storage technologies when considering energy storage integration for a specific NPP.

Alternatively, the costs of storage technologies could be compared by calculating a *levelized cost of storage* (LCOS) for each technology, although this method might not sufficiently address the differences between a technology's power and energy costs [42]. Using each technology's LCOS to compare costs might allow one's comparison to include other factors that affect economics, such as technology lifetime and storage degradation rate. However, a *full cost* comparison was used in this chapter, since the LCOS is a relatively new metric.

5.3.3 Environmental Impact

The greenhouse gas emissions produced by an energy storage system during operation are an example of one of the parameters contributing to a storage technology's environmental impact. Each technology's emissions were categorized as either nonexistent, present in insignificant amounts, or present in significant amounts. Each technology was then assigned a corresponding score of 0, 1, or 2 to represent their GHG emissions, which will help compare the environmental impact of several energy storage options. Since environmental impact can tend to be a more subjective characteristic, calculating an *environmental impact score* allows for a more objective comparison. This methodology has been demonstrated before [43]. The parameters used to index each technology's environmental impact are listed in Table 5.3.

Adding up an energy storage technology's score for each environmental impact parameter yields that technology's *environmental impact score.* For each of the parameters listed in Table 5.3, a higher score corresponds to the

Table 5.3 The Parameters in This Table Were Selected to Describe the Life-Cycle Environmental Impacts of an Energy Storage System and Were Assigned Based on Review of the Available Literature

Environmental Impact	Description
Land and water impact	The footprint of a technology affects the available space for natural flora and fauna. The water footprint of a technology can limit the amount of water available for other purposes. A storage technology's impact was measured to be insignificant, not very significant, somewhat significant, significant, or very significant (i.e., scale from 0 to 4)
Greenhouse gas emissions	The quantity of GHG emissions produced during standard operation. Measured to be absent, present in insignificant amounts, or present in significant amounts (i.e., scale from 0 to 2)
Use of hazardous materials	The use of corrosive or otherwise hazardous materials in an energy storage system's construction can damage the environment if not handled properly during operation and disposal. Measured to be absent, present and recyclable, or present and nonrecyclable (i.e., scale from 0 to 2)

Table 5.3 The Parameters in This Table Were Selected to Describe the Life-Cycle Environmental Impacts of an Energy Storage System and Were Assigned Based on Review of the Available Literature—cont'd

Environmental Impact	Description
Production of hazardous fumes	Dangerous fumes produced by an energy storage system during operation that can negatively affect local wildlife populations. Measured to be either absent or present (i.e., scale from 0 to 1)
Nonenvironmental safety concerns	Some storage technologies present a risk of explosion or other safety concerns (e.g., strong magnetic fields). An energy storage system was measured to have no, minimal, some, or several safety concerns (i.e., scale from 0 to 3)
Resource depletion	The rate at which a nonrenewable natural resource is being depleted to construct a technology. Measured to be insignificant, somewhat significant, or very significant (i.e., scale from 0 to 2)

energy storage system having a more deleterious effect on the environment. Once again, calculating a numerical score in this way helps us compare the energy storage options considered in this chapter more objectively. A more detailed calculation is shown in Appendix D.

5.3.4 Logistical Constraints

The NPP under consideration might have a limited amount of space available to build an energy storage system, which would favor technologies with higher energy densities. Similarly, if the natural formations needed to construct a pumped-storage hydropower (PSH) or compressed-air energy storage (CAES) facility are not available to a potential developer, then these technologies should not be considered as viable solutions. Energy storage technologies that are not compatible with the resources available to an NPP should be eliminated from consideration when selecting favorable solutions.

5.3.5 Regional Policy and Market Conditions

Although policy and market information should not affect the calculation of favorability scores for the considered energy storage systems, the policy and market conditions in the region an NPP is being installed are still important. Since this is a subjective metric, a state's policy stance toward energy storage was ascertained by searching the DOE Global Energy Storage Database for any policies in the state regarding energy storage [44]. Likewise, the regional market conditions were assessed by establishing a *market variability* metric. A market was found to have high variability if the variability of the state's electric grid was measured to be in the 66th percentile of the United States. Variability was evaluated by identifying the percentage of a state's electricity generation from wind and solar energy and comparing this value to the rest of the United States [11]. A market was characterized as having mild variability if its variability lay between the 33rd and 66th percentiles and low variability if it fell within the 33rd percentile. These data are recorded in Appendix B. These conditions would affect nearly all energy storage technologies equally for a given region but could inform an NPP developer's decision to integrate energy storage in the first place.

5.3.6 Application Compatibility

Another important point of comparison for energy storage technologies are the specific benefits the technology can provide to the grid as a source of revenue or increased efficiency for a nuclear power plant. A technology's ability to provide a particular service is dependent on the parameters laid out in Appendix A. As a result, the services offered by electric and thermal energy storage technologies are often different. The storage applications that were considered are defined below. Information was found at the following references [17,24,26,28,45].

- *Energy arbitrage*: Energy arbitrage refers to the process of storing energy when prices are low and selling stored energy when the price of energy is higher. Baseload generators can simulate a flexible output by using energy storage, allowing them to take advantage of changing prices for electricity.
- *Frequency regulation*: Frequency regulation is the practice of balancing momentary differences between generation and demand. This service is required by North American Electric Reliability Corporation (NERC) mandatory reliability standards in an effort to maintain the grid's frequency at 60 Hz. Frequency regulation is typically automated and occurs on a minute-by-minute basis. Energy storage can provide frequency regulation services by discharging when demand exceeds supply and charging when supply exceeds demand.
- *Load following*: While frequency regulation is required to balance momentary differences between the supply and demand of electricity, load following is required to match larger trends in supply and demand.

Load following is characterized as power output that changes every several minutes. As the load changes throughout the day, the generation of electricity must increase to match demand. However, since power is purchased hourly, load-following services are needed to follow the load between auctions.

- *Voltage support*: Grid operators must maintain stable voltage levels in the transmission and distribution system. However, reactance produced by electronic equipment connected to the grid threatens to cause unacceptable voltage fluctuations. Reactive power must be injected to the grid to offset these fluctuations. Residential PV systems are a growing source of reactance on the grid.
- *Spinning, nonspinning, and supplemental reserves*: Reserves are needed to supply power to the grid in case any part of the supply suddenly becomes unavailable. In the United States, 15%–20% of the normal electricity supply capacity is usually available in reserves at any time. However, the reserves that are available to the grid cannot all respond to an outage immediately. Spinning reserves are generators that are online, but not supplying power to the grid; spinning reserves can respond to an outage within 10 s to 10 min. Nonspinning reserves are generators that are off-line but can respond within 10 min. Nonspinning reserves can also be power plants that are not operating at full capacity and can ramp up in response to an outage. Supplemental reserves are the slowest to respond and can come online within approximately 1 h, depending on the type of power plant. Energy storage technologies can often simulate spinning reserves due to quick-response times.
- *Black start capabilities*: Black start capabilities are needed to energize the grid when the grid collapses and all other reserve capacity fails to back up the grid. Black start capabilities can provide power to consumers and restart power plants without drawing power from the grid. Energy storage technologies are well suited to provide this service to the grid.
- *Variable supply resource integration*: Energy storage can be used to optimize the output from VRES to increase the value of the transmitted electricity. In particular, energy storage can provide two valuable services to renewable energy sources. Energy storage can be used for capacity firming or enabling the use of an intermittent supply as a constant power source. In this chapter, this parameter refers to an energy storage technology's ability to assist with the integration of VRES with the electric grid.

- *Process heat applications*: The heat from a nuclear power plant could be stored with a thermal energy storage technology and used to power an external process that requires heat. For example, energy storage technologies can be used in combined heat and power plants to temporally align the consumer demand for electricity and heat.
- *Seasonal storage*: Energy storage technologies can be used to store energy for long periods of time to compensate for seasonal changes in supply and demand. For example, a thermal energy storage technology can be used to store heat in the summer to be used in the winter when this resource becomes more necessary.

5.3.7 Favorability Analysis

Finally, to determine a technology's overall compatibility with a specific application, the environmental impact, technical maturity, and economic feasibility for each energy storage option should be compared. This could be done by adding up each of the discussed numerical scores (i.e., TRL, full cost or economic feasibility, and environmental impact) and weighting these scores according to the user's individual preferences. A similar methodology was employed in this chapter when comparing energy storage options, as shown below:

$$F = EI \cdot \overset{[0,1]}{W_1} + EF \cdot \overset{[0,1]}{W_2} + T \cdot \overset{[0,1]}{W_3} \tag{5.1}$$

where W represents the chosen weighting factors, EI is the environmental impact score, EF is the economic feasibility score, T is the technical maturity, and F represents the overall favorability score. The weighting factors in this equation allow the user to change the relative importance of each component score to match the individual characteristics of the NPP under consideration (by assigning a value between zero and one). Appendix D contains additional information regarding the calculation of the terms in Eq. (5.1) If a single commercial application is being analyzed, then the technologies can also be sorted by whether the storage technologies are fully compatible, somewhat compatible, or incompatible with the required service (these data are recorded in Appendix A). Thus, the technology with the highest favorability score that is also fully compatible with the required service would be recommended for the application under consideration.

However, in addition to the weighted sum in Eq. (5.1), some energy storage options should be eliminated from consideration due to several constraints. Budget constraints determine whether the cost of a technology option exceed the user's budget for installing an energy storage system. Additionally, logistic constraints consider whether a technology has sufficient space and has access to any required geographic features and whether a technology has sufficient mobility, if a mobile system is desired.

5.4 CASE STUDIES

Two case studies are discussed in detail to show how energy storage technologies might be compared for real-world applications.

5.4.1 Case Study #1: Pumped Storage Hydropower in France

France presents a unique scenario for energy storage and nuclear power due to the country's high concentration of electricity generation from nuclear power. Hydroelectric facilities in France serve an important secondary purpose by operating as an inexpensive and flexible form of energy storage. This energy storage capacity is critical since the start-up time of a typical nuclear reactor can be up to 40 h, while almost 15,000 MW of hydroelectric capacity can be brought online in a matter of minutes [46]. Although both run-of-river- and reservoir-type hydroelectric facilities can operate as energy storage systems, the type of hydroelectric facility considered in this analysis is pumped-storage hydropower. These facilities can be used to pump water to a higher elevation during periods of low demand so that this water can be used to generate additional electricity to meet peak load. PSH facilities make up approximately 16% of hydropower capacity in France, or around 4–5 GW. On a normal day, this energy storage capacity is used to provide about 4 h of additional generation during periods of high consumption, as displayed in Fig. 5.2 [46].

For this case study, it was assumed that an energy storage developer was hired to address the French government's concerns and provide additional flexibility for the nation's large nuclear power fleet. Table 5.4 provides parameters that constrain the case study for France. A timeline of 1–5 years was chosen because technical maturity is critical to this simulation, considering the actual energy storage deployment in France occurred decades ago.

To compare energy storage options for this case study, the weighting factors for technical maturity, economic feasibility, and environmental impact were set to 1, 0.8, and 0.5, respectively. Note that the selection of these weighting factors is mostly arbitrary and could be changed by another interested user. The resulting simple calculation yields the following energy

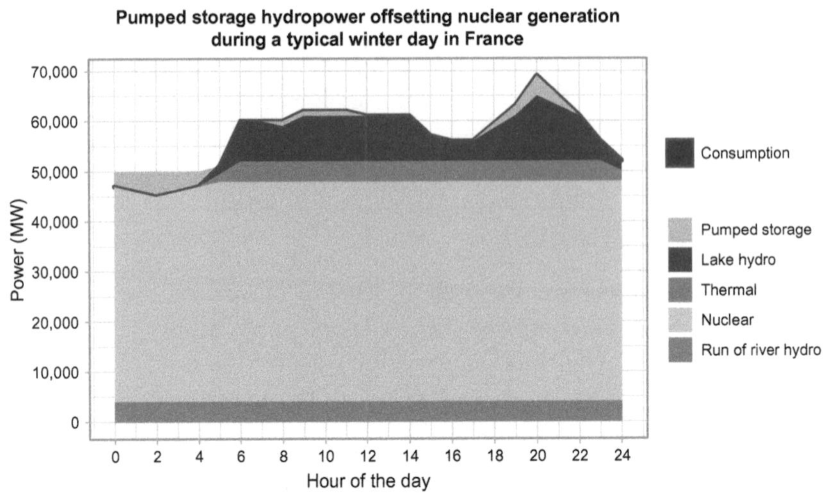

Fig. 5.2 France uses pumped-storage hydropower powered by nuclear during times of low demand to provide additional generation capacity at a later time to meet peak demand [46].

Table 5.4 Parameters Used to Describe the Situation of Using Energy Storage to Offset Electricity Generation From Nuclear Power in France

Parameter	Value
Required storage capacity	16 GWh
Required power	4 GW
Available budget	$40 million
Grid-scale service	Energy arbitrage
Timeline	1–5 years
Available geographic features	All

storage options in order of recommendation: (1) molten salts, (2) hot and cold water, (3) compressed-air energy storage (CAES), (4) lead-acid batteries, (5) sodium-sulfur batteries, and (6) PSH. The result of PSH being listed in sixth place is at odds with the fact that the French chose to install PSH as their primary source of energy storage.

Additional investigation is required to explain this discrepancy. The two thermal energy storage systems on this list (molten salts and heated water) can be removed from consideration, since thermal energy cannot be transported efficiently across large distances relative to the ability to transport electricity [47]. Lead-acid and sodium-sulfur batteries are potential solutions since electricity can be transported efficiently, but these technologies had not yet been developed for large-scale installations during the implementation of the French system. Even today, the largest battery installations are no larger than 400 MWh [44]. As a result, batteries would only have been feasible as a distributed solution, which mitigates the economies of scale that can be achieved with centralized facilities.

This leaves only CAES as a superior recommendation over PSH. However, CAES facilities have only been successfully deployed a handful of times and require unique geologic conditions [24]. It is unlikely that the French government would have considered CAES to be a viable option at the time. The remaining preferred technology is PSH, which was the energy storage system chosen by the French. This scenario demonstrates how it is difficult to recommend a clear-cut winner when comparing energy storage options, but this chapter intends to provide the information and decision-making framework to select a solution from a group of viable options.

5.4.2 Case Study #2: Advanced Nuclear Power Plant in the United States

To provide an accurate picture of how favorable energy storage technologies might be selected for an advanced NPP built in the United States, the specifications of a single reactor at the Vogtle 3 and 4 site in Waynesboro, GA, were considered, where two new reactor units are being built by Southern Company [48]. Each of these reactors will utilize a Generation III+ reactor design, the Westinghouse AP1000, in compliance with the Nuclear Regulatory Commission's (NRC) stricter safety standards implemented in response to the Fukushima disaster [19]. Each reactor at this site has an electricity generation capacity of 1117 MW [48]. Ideally, the chosen energy storage system would be able to offset the entire generation capacity of the advanced NPP when electricity prices are low, but a facility of this size is unprecedented for most of the considered energy storage systems (e.g., batteries).

The Westinghouse AP1000 reactor is also capable of some load following, so the energy storage system is only needed to supplement the advanced NPP's flexibility and provide additional income. In particular, the AP1000 reactor is capable of load following for up to 90% of its fuel cycle with a ±5%/min ramp rate and a ±10%/min step load change. The AP1000 is designed to cycle its power level from 100% to 50% and back to 100% when load following. This reactor is also designed to perform this fast cycling without generating excessive waste water or generating severe axial xenon oscillations, which improves the reactor's recovery time after cycling (i.e., 20%

power step increase or decrease within 20 min) [20]. Therefore, the maximum power output of the energy storage system was set at 30 MW. This aligns well with existing energy storage units and the available PSH resources from nonpowered dams. As with the first case study, the energy storage capacity of the system was chosen by sizing the system to offset 4 h of electricity generated during a period of low demand.

According to the current construction schedule for Vogtle 3 and 4, the reactor units are expected to be deployed in 2021 and 2022, respectively. Therefore, the installation of an accompanying energy storage system was projected to occur 6–10 years from now to account for the storage technology's construction timeline. Due to the plant's location in Waynesboro, GA, the energy storage system should have no significant space or weight constraints, since there is sufficient open land nearby. However, the footprint of the energy storage facility was still limited to 1% of the advanced NPP's footprint, which is approximately 2000 acres, to keep the size of the chosen energy storage facility within reasonable limits [48]. The developer's available budget for an energy storage system was estimated to be 1 million dollars based on the size of the facility (~0.1% of the total budget). This is probably reasonable considering Southern Company's budget for Vogtle 3 and 4 of approximately 20 billion dollars [49]. This budget was large enough to fund most of the energy storage options considered in this work and meet the requirements of the case study as detailed in Table 5.5. In future investigations, a much larger budget might be justified if the economic value of an integrated energy storage system has been successfully demonstrated.

Table 5.5 Parameters Used to Describe a Potential Energy Storage System for Plant Vogtle, Units 3 and 4

Parameter	Value
Required storage capacity	120 MWh
Required power	30 MW
Available budget	$1 million
Timeline	6–10 years
Available geographic features	Underground aquifer, elevation change, water source
Grid-scale service	Energy arbitrage, frequency regulation, load following, spinning reserves, nonspinning reserves, voltage support

The compatibility of the energy storage options considered in this chapter with the geography of the area surrounding Waynesboro, GA, should also be evaluated. First, the location of Vogtle 3 and 4 does not appear to be compatible with CAES, due to the lack of geologic formations such as salt and rock caverns [50]. However, based on an assessment of nonpowered dams in the United States, it appears that there is potential for PSH development as an energy storage system [51]. A map displaying possible locations for further PSH development is displayed in Fig. 5.3. This map indicates that 1–30 MW of additional energy generation is available near Waynesboro, GA, which coincides with the chosen power output of the potential energy storage system.

Underground thermal energy systems (UTES) also have specific geographic requirements similar to CAES and PSH facilities. This technology makes use of existing underground aquifers or lakes, geothermal resources, or sufficiently large caverns to store thermal energy. Data were gathered from a variety of sources to determine the availability of these resources nearby Waynesboro, GA. First, according to a map acquired from the US

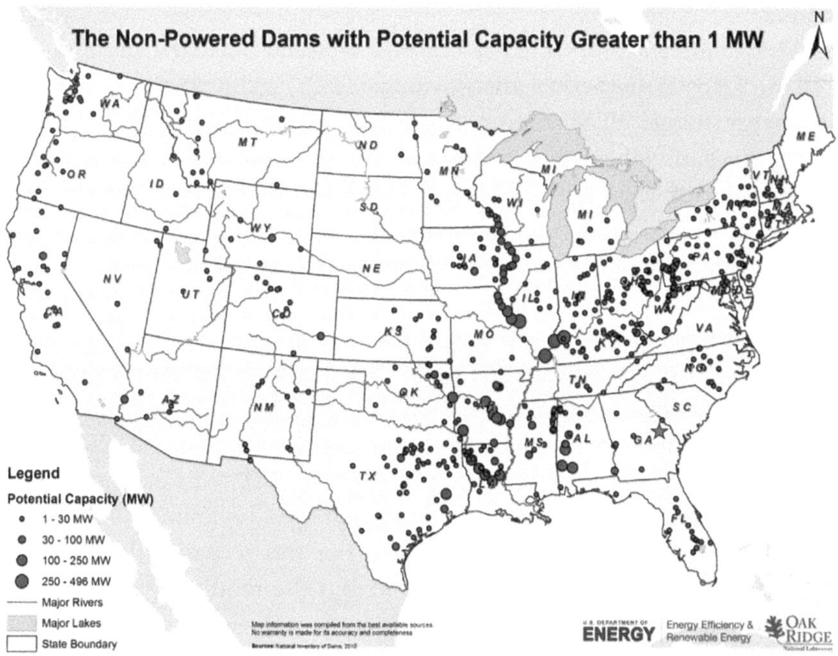

Fig. 5.3 Locations of nonpowered dams in the United States for potential new hydroelectric capacity [51].

Geological Survey, it appears that the southeastern coastal plain aquifer system runs relatively close to the sites of the Vogtle 3 and 4 reactors [52]. Therefore, for the sake of this simplified case study, UTES facilities were considered as viable energy storage options for Vogtle 3 and 4. Similar maps were obtained to assess the availability of geothermal resources and cave systems in the United States, but this information revealed that the geologic resources in the area are not adequate [53,54].

Table 5.5 summarizes the specifications that were used in this case study to simulate the requirements for energy storage integration with Vogtle 3 and 4. The grid-scale services listed in Table 5.5 correspond to the ancillary services that have been defined by Southern Company [55]. However, the results displayed in Fig. 5.4 are given in a *compatibility-agnostic* format. Since the application of an integrated energy storage system is highly dependent on the intentions of the developer installing the energy storage system, it is difficult to assume how energy storage will be used on the grid. The parameters displayed in Table 5.5 are specific to an energy storage system primarily used for shifting energy time-of-use, since it is likely that any integrated technology will perform this service to increase the flexibility and market power of the NPP.

The parameter values listed in Table 5.5 were altered to test an additional scenario focused on thermal energy storage (TES) technologies. With electric energy storage (EES) systems, it is simple to offset 4 h of energy generated during one part of the day by storing it and discharging the energy directly to the grid during a period of the day with more highly priced electricity. However, a TES system must be equipped with the means to generate electricity for the system to similarly discharge stored energy to the grid. Transferring 4 h of energy to another 4 h period could require the developer to install additional turbines at the power plant along with the additional infrastructure required to facilitate the steam cycle. Instead, the same 4 h of low-priced energy could be stored and then discharged over the remaining 20 h of the day to supplement the plant's output by a small amount. With TES, the advanced NPP could avoid selling electricity below its marginal costs, increase the amount of electricity sold to the grid, and minimize the amount of additional generation capacity needed. For this scenario, the required energy storage capacity remains the same, but the required power output was reduced to 6 MW.

Assume that the EES and TES system options described above are defined as Scenario A and Scenario B, respectively. The results for EES and TES systems were averaged to obtain each technology's overall

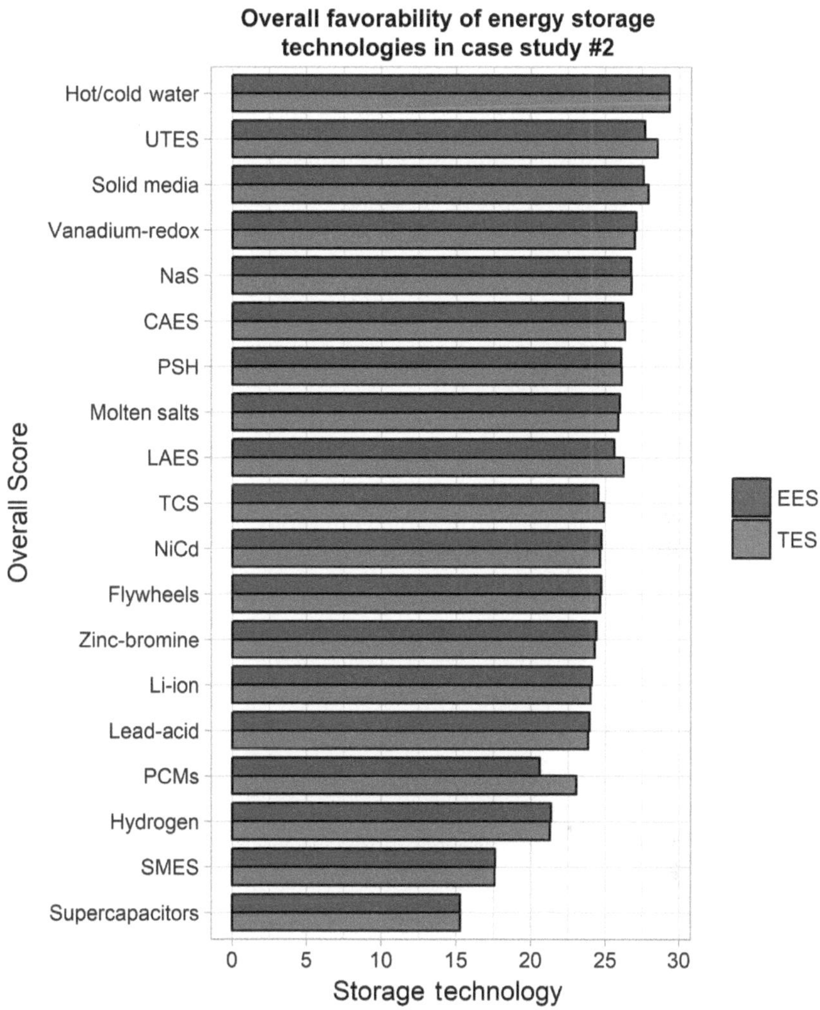

Fig. 5.4 Energy storage technologies ranked by overall favorability scores to reveal the most appropriate options.

favorability score. The energy storage technologies considered in this chapter were ranked according to these favorability scores, which were calculated using the parameters recorded in Table 5.5. Unlike Case Study #1, the environmental impact, technical maturity, and economic feasibility scores for each energy storage option were weighted equally in Case Study #2. Interestingly, there were only very slight differences between the two scenarios analyzing EES and TES systems for Case Study #2. In Scenario A (EES), PSH was ranked more favorably than liquid air energy storage (LAES), and CAES was favored over UTES, while in Scenario B (TES), the opposite was true. The averaged favorability scores for each energy storage option are displayed in Fig. 5.4. Note that in Fig. 5.4, the favorability scores have been reduced to a scale between 0 and 30. This could be done by dividing each score by the highest overall value and then multiplying each score by 30 to recalibrate. In this chart, phase change materials (PCMs) and superconducting magnetic energy storage (SMES) are represented by their respective acronyms.

5.5 CLOSING SUMMARY

This chapter describes several energy storage technologies compatible with NPPs and critically compares the characteristics of these energy storage options to determine the most promising technology for specific plants. This comparison includes analyzing the environmental impact, technical maturity, economic feasibility, logistic constraints, and policy and market considerations for specific grid-scale applications using a wide range of storage technologies. Two case studies were analyzed to illuminate a potential framework for objectively comparing technology options. While the scenario considered in Case Study #1 is useful for validation, Case Study #2 provides a more detailed view of which energy storage systems might hold the most potential for successful integration alongside a newly constructed advanced NPP. Since the storage technologies in Fig. 5.4 are ranked by their favorability scores with no consideration given to their compatibility with various grid-scale services, some filtering is still required before one might be able to identify the best technology for their application.

According to the results in Fig. 5.4, hot and cold water storage was identified as the most favorable energy storage system for the scenario presented in Case Study #2. Hot and cold water storage is a mature technology, although it has primarily been used for shifting thermal energy throughout the day for building heating and cooling. This technology is also inexpensive and has a minimal environmental impact; however, additional hardware, including extra turbines, might be required to convert the thermal energy into electricity if the NPP under consideration plans to continue operating at full capacity while the TES system is used for energy arbitrage.

A distinction can be made between the group of energy storage systems with favorability scores >25 and the rest of the considered technologies. In this more favorable group, we see a high concentration of thermal energy storage technologies and representation from both "bulk energy storage" technologies (CAES and PSH). This is most likely due to the low environmental impact and costs associated with these technologies. A couple of batteries, sodium-sulfur and vanadium redox, are also featured in the upper tier of technologies considered in Case Study #2. A few storage technologies that have seen widespread deployment in recent years, including lithium-ion batteries and flywheels, fall much lower on the list of preferred technologies for the case studies considered. Lithium-ion batteries and flywheels are uniquely suited for grid-level applications, like supplying frequency regulation and voltage support. When considering an energy storage system for installation alongside an NPP, however, these technologies are either too expensive for energy arbitrage or environmentally impactful compared with the other available options. The relatively low performance of lithium-ion batteries in Case Study #2 highlights the importance of identifying the specific needs of one's application to properly rank the available energy storage options.

Lithium-ion batteries, flywheels, and other similar EES technologies are also popular due to their compatibility with renewables. The quick charging and discharging characteristics and minimal geologic and environmental requirements offered by electrochemical batteries and flywheels match well with the intermittent nature of remote wind and solar energy. As a baseload generator, NPPs have different needs that might make energy storage solutions like PSH, CAES, and TES more suitable since they excel at storing large amounts of energy efficiently for a span of a few hours in the case of TES or several weeks with PSH and CAES. Vanadium redox and sodium-sulfur batteries also show effectiveness in long-term energy storage when compared with most other EES technologies. With additional commercial deployments of electrochemical batteries, the technical and economic competitiveness might improve to the point that lithium-ion batteries could be considered viable for complementing an NPP.

APPENDIX A: PERFORMANCE METRICS FOR THE CONSIDERED ENERGY STORAGE TECHNOLOGIES

These metrics were first recorded in *An Evaluation of Energy Storage Options for Nuclear Power* by Coleman et al. and are reproduced here. Further explanation for these metrics can be found in Coleman's manuscript [56].

1. Mechanical Energy Storage

Table A.1 Performance Metrics for Mechanical Energy Storage

Storage Technology	Energy Capacity (MWh)	Power Capacity (MW)	Energy Capacity Cost ($/kWh)	Power Capacity Cost ($/kW)	Discharge Time
PSH	500–8000 [30]	100–5000 [28]	5–100 [30]	2000–4000 [30]	6–10 h [30]
CAES	580 and 2860 [30]	110 and 290 [30]	2–120 [28]	500–1500 [24,30]	8–20 h [30]
Flywheels	0.0005–0.025 per unit, five total [26,57]	0.1–1.65 per unit, 20 total [26,57]	1000–5000 [30]	250–350 [30]	0–0.25 h [28]

Storage Technology	Response Time	Storage Degradation Rate (%/Day)	Energy Density (kWh/m^3)	Power Density (kW/m^3)	Specific Energy (Wh/kg)
PSH	Minutes [28]	Very small [30]	0.5–1.5 [30]	0.5–1.5 [30]	0.5–1.5 [30]
CAES	Seconds to minutes [28]	Small [30]	2–6 [30]	0.5–2 [30]	30–60 [30]
Flywheels	Seconds [28]	20% per h [30]	20–80 [30]	1000–2000 [30]	10–30 [30]

Storage Technology	Specific Power (W/kg)	Round-Trip Efficiency	Cycle Life (Cycles)	Technology Lifetime (Years)	O&M Costs ($/kW/Year)
PSH	–	76%–85% [26]	10,000–30,000 [28]	50–60 [26]	~3 [30]
CAES	–	~70% [30]	8000–12,000 [28]	20–40 [28]	19–25 [30]
Flywheels	400–1500 [30]	90%–95% [28]	20,000–100,000 [26,28]	15–20 [26,28]	~20 [30]

Storage Technology	Technology Readiness Level	Storage Output Temperature (°C)
PSH	9 [26]	–
CAES	9 [26]	–
Flywheels	7 [26]	–

Table A.2 Environmental Impacts for Mechanical Energy Storage

Environmental Impact	PSH	CAES	Flywheels
Land and water impact	Very significant	Somewhat significant	Insignificant
Emissions produced during operation	Yes, but not very significant	Yes	None
Hazardous materials	None	None	None
Hazardous fumes	None	None	None
Short-term safety concerns	Some	Some	Some
Resource depletion	Insignificant	Insignificant	Insignificant
Geographic requirements	Yes	Yes	None

Table A.3 Compatible Applications for Mechanical Energy Storage

Service	PSH	CAES	Flywheels
Energy arbitrage	Compatible	Compatible	Incompatible
Frequency regulation	Somewhat compatible	Somewhat compatible	Compatible
Load following	Compatible	Compatible	Somewhat compatible
Voltage support	Incompatible	Incompatible	Compatible
Spinning reserves	Somewhat compatible	Somewhat compatible	Compatible
Nonspinning and supp reserves	Compatible	Compatible	Incompatible
Black start	Compatible	Compatible	Incompatible
VSR integration	Compatible	Compatible	Compatible
Seasonal storage	Compatible	Compatible	Incompatible
Process heat applications	Incompatible	Incompatible	Incompatible

2. Electrical Energy Storage

Table A.4 Performance Parameters for Electrical Energy Storage Systems

Storage Technology	Energy Capacity (MWh)	Power Capacity (MW)	Energy Capacity Cost ($/kWh)	Power Capacity Cost ($/kW)	Discharge Time
Supercapacitors	0.0005 [30]	0–0.3 [28]	10,000 [58]	130–515 [24]	Milliseconds to 1 h [30]
SMES	0.001–0.015 [30]	0.1–10 [30]	1000–10,000 [28,30]	200–300 [30]	Milliseconds to seconds [30]

Continued

Table A.4 Performance Parameters for Electrical Energy Storage Systems—cont'd

Storage Technology	Response Time	Storage Degradation Rate (%/Day)	Energy Density (kWh/m^3)	Power Density (kW/m^3)	Specific Energy (Wh/kg)
Supercapacitors	Milliseconds, <1/4 cycle [28]	20%–40% [30]	0.01–1 [59]	200–10,000 [60]	2.5–15 [30]
SMES	Milliseconds, <1/4 cycle [28]	10%–15% [30]	0.2–2.5 [30]	1000–4000 [30]	0.5–5 [30]

Storage Technology	Specific Power (W/kg)	Round-Trip Efficiency	Cycle Life (Cycles)	Technology Lifetime (Years)	O&M Costs ($/kW/Year)
Supercapacitors	500–5000 [30]	90%–95% [28]	100,000 + [28]	10–30 [30]	~6 [30]
SMES	500–2000 [30]	95%–98% [28]	100,000 + [28,30]	20–30 [28,30]	18.5 [30]

Storage Technology	Technology Readiness Level	Storage Output Temperature (°C)
Supercapacitors	5 [24]	–
SMES	5 [24]	–

Table A.5 Environmental Impacts for Electrical Energy Storage Systems

Environmental Impact	Supercapacitors	SMES
Land and water impact	Insignificant	Insignificant
Emissions produced during operation	None	None
Hazardous materials	None	None
Hazardous fumes	None	None
Short-term safety concerns	Minimal	Several
Resource depletion	Somewhat significant	Somewhat significant
Geographic requirements	None	None

Table A.6 Compatible Applications for Electrical Energy Storage Systems

Service	Supercapacitors	SMES
Energy arbitrage	Incompatible	Incompatible
Frequency regulation	Compatible	Compatible
Load following	Compatible	Compatible
Voltage support	Compatible	Compatible

Table A.6 Compatible Applications for Electrical Energy Storage Systems—cont'd

Service	Supercapacitors	SMES
Spinning reserves	Somewhat compatible	Somewhat compatible
Nonspinning and supp reserves	Incompatible	Incompatible
Black start	Incompatible	Incompatible
VSR integration	Compatible	Incompatible
Seasonal storage	Incompatible	Incompatible
Process heat applications	Incompatible	Incompatible

3. Electrochemical Energy Storage—Conventional Batteries

Table A.7 Performance Parameters for Conventional Battery Systems

Storage Technology	Energy Capacity (MWh)	Power Capacity (MW)	Energy Capacity Cost ($/kWh)	Power Capacity Cost ($/kW)	Discharge Time
Lithium-ion	0.25–25 [57]	0.005–50 [26]	600–2500 [30]	1200–4000 [30]	Minutes to hours [30]
NaS	~300 [57]	~50 [57]	300–500 [30]	1000–3000 [30]	Seconds to hours [30]
Lead-acid	0.001–40 [30]	~0–20 [28,30]	200–400 [30]	300–600 [28,30]	Seconds to hours [30]
NiCd	~6.75 [30]	~0–40 [28,30]	800–1500 [30]	500–1500 [30]	Seconds to hours [30]

Storage Technology	Response Time	Storage Degradation Rate (%/Day)	Energy Density (kWh/m^3)	Power Density (kW/m^3)	Specific Energy (Wh/kg)
Lithium-ion	Milliseconds, <1/4 cycle [30]	0.1%–0.3% [30]	200–500 [30]	30–300 [59]	75–200 [30]
NaS	Milliseconds, <1/4 cycle [28]	Almost zero [30]	150–250 [30]	140–180 [30]	150–240 [30]
Lead-acid	Milliseconds, <1/4 cycle [28]	0.1%–0.3% [30]	50–80 [30]	10–400 [30]	30–50 [30]
NiCd	Milliseconds, <1/4 cycle [28]	0.2%–0.6% [30]	60–150 [30]	80–600 [30]	50–75 [30]

Continued

Table A.7 Performance Parameters for Conventional Battery Systems—cont'd

Storage Technology	Specific Power (W/kg)	Round-Trip Efficiency	Cycle Life (Cycles)	Technology Lifetime (Years)	O&M Costs ($/kW/Year)
Lithium-ion	750–1250 [61]	75%–90% [30]	~3000 at 80% DOD [30]	5–15 [28,30]	10 [62]
NaS	150–230 [30]	75%–90% [28,30]	2500–4500 [28,30]	10–15 [28,30]	80 [30]
Lead-acid	75–300 [30]	70%–80% [30]	500–1000 [28,30]	5–15 [28,30]	50 [30]
NiCd	150–300 [30]	60%–70% [30]	2000–2500 [28,30]	10–20 [28,30]	20 [30]

Storage Technology	Technology Readiness Level	Storage Output Temperature (°C)
Lithium-ion	9 [63]	–
NaS	8 [26]	–
Lead-acid	9 [57]	–
NiCd	7 [30]	–

Table A.8 Environmental Impacts of Conventional Battery Systems

Environmental Impact	Lithium-Ion	NaS	Lead-Acid	NiCd
Land and water impact	Insignificant	Insignificant	Insignificant	Insignificant
Emissions produced during operation	None	None	None	None
Hazardous materials	Yes	Yes, recyclable	Yes, recyclable	Yes
Hazardous fumes	None	None	Yes	None
Short-term safety concerns	Several	Some	Several	None
Resource depletion	Somewhat significant	Somewhat significant	Very significant	Very significant
Geographic requirements	None	None	None	None

Table A.9 Compatible Applications for Conventional Battery Systems

Service	Lithium-Ion	NaS	Lead-Acid	NiCd
Energy arbitrage	Somewhat compatible	Compatible	Compatible	Somewhat compatible
Frequency regulation	Compatible	Compatible	Compatible	Incompatible
Load following	Somewhat compatible	Compatible	Somewhat compatible	Incompatible
Voltage support	Compatible	Compatible	Compatible	Incompatible
Spinning reserves	Somewhat compatible	Compatible	Compatible	Compatible
Nonspinning and supp reserves	Incompatible	Compatible	Somewhat compatible	Incompatible
Black start	Somewhat compatible	Compatible	Compatible	Compatible
VSR integration	Compatible	Compatible	Compatible	Incompatible
Seasonal storage	Incompatible	Incompatible	Incompatible	Incompatible
Process heat applications	Incompatible	Incompatible	Incompatible	Incompatible

4. Electrochemical Energy Storage—Flow Batteries

Table A.10 Performance Parameters for Flow Battery Systems

Storage Technology	Energy Capacity (MWh)	Power Capacity (MW)	Energy Capacity Cost ($/kWh)	Power Capacity Cost ($/kW)	Discharge Time
ZnBr	~250 [57]	~50 [57]	150–1000 [30]	700–2500 [30]	Seconds to ~10 h [30]
VRB	~250 [57]	~50 [57]	150–1000 [30]	600–1500 [30]	Seconds to 24 + h [30]

Storage Technology	Response Time	Storage Degradation Rate (%/Day)	Energy Density (kWh/m^3)	Power Density (kW/m^3)	Specific Energy (Wh/kg)
ZnBr	Milliseconds, < 1/4 cycle [28]	Small, almost zero when electrolyte stored separately [30]	30–60 [30]	<25 [30]	30–50 [30]
VRB	Milliseconds, <1/4 cycle [28]	Small, almost zero when electrolyte stored separately [30]	25–35 [30]	<2 [30]	10–30 [30]

Continued

Table A.10 Performance Parameters for Flow Battery Systems—cont'd

Storage Technology	Specific Power (W/kg)	Round-Trip Efficiency	Cycle Life (Cycles)	Technology Lifetime (Years)	O&M Costs ($/kW/Year)
ZnBr	~100 [30]	65%–75% [30]	2000+ [28,30]	5–10 [28,30]	–
VRB	~166 [30]	65–75% [30]	12,000+ [28,30]	5–10 [28,30]	70 [30]

Storage Technology	Technology Readiness Level	Storage Output Temperature (°C)
ZnBr	6 [26]	–
VRB	7 [26]	–

Table A.11 Environmental Impacts of Flow Battery Systems

Environmental Impact	ZnBr	VRB
Land and water impact	Insignificant	Insignificant
Emissions produced during operation	None	None
Hazardous materials	Yes, recyclable	Yes, recyclable
Hazardous fumes	None	None
Short-term safety concerns	Minimal	None
Resource depletion	Very significant	Somewhat significant
Geographic requirements	None	None

Table A.12 Compatible Applications for Flow Battery Systems

Service	ZnBr	VRB
Energy arbitrage	Somewhat compatible	Somewhat compatible
Frequency regulation	Compatible	Compatible
Load following	Compatible	Compatible
Voltage support	Compatible	Compatible
Spinning reserves	Compatible	Compatible
Nonspinning and supp reserves	Compatible	Compatible
Black start	Compatible	Compatible
VSR integration	Compatible	Compatible
Seasonal storage	Somewhat compatible	Somewhat compatible
Process heat applications	Incompatible	Incompatible

5. Chemical Energy Storage

Table A.13 Performance Parameters for Chemical Energy Storage Systems

Storage Technology	Energy Capacity (MWh)	Power Capacity (MW)	Energy Capacity Cost ($/kWh)	Power Capacity Cost ($/kW)	Discharge Time
Hydrogen fuel cell	1000–1000,000 (underground cavern) [32]	0–50 [30]	15 [30]	1500–3000 [30]	Seconds to 24+ h [30]

Storage Technology	Response Time	Storage Degradation Rate (%/Day)	Energy Density (kWh/m^3)	Power Density (kW/m^3)	Specific Energy (Wh/kg)
Hydrogen fuel cell	Seconds, <1/4 cycle [28,30]	Almost zero [30]	500–3000 [30]	500+ [30]	800–10,000 [30]

Storage Technology	Specific Power (W/kg)	Round-Trip Efficiency	Cycle Life (Cycles)	Technology Lifetime (Years)	O&M Costs ($/kW/Year)
Hydrogen fuel cell	500–800 [30]	20%–30% [32]	1000+ (fuel cell) [28,30]	5–15 [28, 30]	0.0019–0.0153 $/kW [30]

Storage Technology	Technology Readiness Level	Storage Output Temperature (°C)
Hydrogen fuel cell	6 [29]	–

Table A.14 Environmental Impacts of Chemical Energy Storage Systems

Environmental Impact	Hydrogen
Land and water impact	Significant
Emissions produced during operation	None
Hazardous materials	Yes
Hazardous fumes	None
Short-term safety concerns	Some
Resource depletion	Insignificant
Geographic requirements	Yes

Table A.15 Compatible Applications for Chemical Energy Storage Systems

Service	Hydrogen
Energy arbitrage	Somewhat compatible
Frequency regulation	Incompatible
Load following	Somewhat compatible
Voltage support	Incompatible
Spinning reserves	Somewhat compatible
Nonspinning and supp reserves	Compatible
Black start	Somewhat compatible
VSR integration	Incompatible
Seasonal storage	Compatible
Process heat applications	Incompatible

6. Thermal Energy Storage—Sensible Heat

Table A.16 Performance Parameters for Sensible Thermal Energy Storage Systems

Storage Technology	Energy Capacity (MWh)	Power Capacity (MW)	Energy Capacity Cost (\$/kWh)	Power Capacity Cost (\$/kW)	Discharge Time
UTES	~3900 [64]	–	~0.055 [64]	3400–4500 [24]	–
Hot and cold water (storage tanks)	10–2000 [24]	–	0.1–10 [33]	300–600 [24]	Minutes to hours [65]
Solid media (concrete)	>1100 [35]	–	~40 [35]	500–3000 [24]	~1 day [35]

Storage Technology	Response Time	Storage Degradation Rate (%/Day)	Energy Density (kWh/m^3)	Power Density (kW/m^3)	Specific Energy (Wh/kg)
UTES	–	Almost zero [33]	20–30 [65]	–	–
Hot and cold water (storage tanks)	Seconds to hours [65]	Almost zero [33]	20–30 [65]	–	–
Solid media (concrete)	–	–	~22 [35]	–	~5 [66]

Table A.16 Performance Parameters for Sensible Thermal Energy Storage Systems—cont'd

Storage Technology	Specific Power (W/kg)	Round-Trip Efficiency	Cycle Life (Cycles)	Technology Lifetime (Years)	O&M Costs ($/kW/Year)
UTES	–	50%–90% [24]	–	–	–
Hot and cold water (storage tanks)	–	50%–90% [24]	–	10–30+ [65]	–
Solid media (concrete)	–	50%–90% [24]	–	>2 [35]	–

Storage Technology	Technology Readiness Level	Storage Output Temperature (°C)
UTES	8 [24]	< 250 [24]
Hot and cold water (storage tanks)	7 [24]	95–98 or 120–130 (pressurized) [24]
Solid media (concrete)	6 [35]	350 [35]

Table A.17 Environmental Impacts of Sensible Thermal Energy Storage Systems

Environmental Impact	UTES	Hot and Cold Water (Storage Tanks)	Solid Media (Concrete)
Land and water impact	Significant	Insignificant	Insignificant
Emissions produced during operation	None	None	None
Hazardous materials	None	None	None
Hazardous fumes	None	None	None
Short-term safety concerns	None	None	None
Resource depletion	Insignificant	Insignificant	Insignificant
Geographic requirements	Yes	None	None

Table A.18 Applications Compatible With Sensible Thermal Energy Storage Systems

Service	UTES	Hot and Cold Water (Storage Tanks)	Solid Media (Concrete)
Energy arbitrage	Incompatible	Compatible	Compatible
Frequency regulation	Incompatible	Incompatible	Incompatible
Load following	Incompatible	Incompatible	Incompatible

Continued

Table A.18 Applications Compatible With Sensible Thermal Energy Storage Systems—cont'd

Service	UTES	Hot and Cold Water (Storage Tanks)	Solid Media (Concrete)
Voltage support	Incompatible	Incompatible	Incompatible
Spinning reserves	Incompatible	Incompatible	Incompatible
Nonspinning and supp reserves	Incompatible	Compatible	Compatible
Black start	Incompatible	Incompatible	Incompatible
VSR integration	Incompatible	Compatible	Compatible
Seasonal storage	Compatible	Compatible	Incompatible
Process heat applications	Compatible	Compatible	Compatible

7. Thermal Energy Storage—Latent Heat

Table A.19 Performance Parameters for Latent Thermal Energy Storage Systems

Storage Technology	Energy Capacity (MWh)	Power Capacity (MW)	Energy Capacity Cost ($/kWh)	Power Capacity Cost ($/kW)	Discharge Time
TCS	–	–	8–100 [33]	1000–3000 [24]	1–24+ h [30]
Molten salts	~350 [38]	–	5–10 [66]	400–700 [24]	–
LAES	20–1000 [39]	–	260–530 [30]	900–1900 [30]	Several hours [30]
PCMs	–	–	10–50 [33]	6000–15,000 [33]	–

Storage Technology	Response Time	Storage Degradation Rate (%/Day)	Energy Density (kWh/m^3)	Power Density (kW/m^3)	Specific Energy (Wh/kg)
TCS	–	Almost zero [30]	140–830 [37]	–	–
Molten salts	–	Very small [24]	170–420 [66]	–	80–190 [66]
LAES	Minutes [30]	<0.2% [39]	–	–	100–140 [39]
PCMs	–	Almost zero [33]	100 [33]	–	–

Table A.19 Performance Parameters for Latent Thermal Energy Storage Systems—cont'd

Storage Technology	Specific Power (W/kg)	Round-Trip Efficiency	Cycle Life (Cycles)	Technology Lifetime (Years)	O&M Costs ($/kW/Year)
TCS	–	80%–99% [24]	–	10–30 + [33]	–
Molten salts	–	40%–93% [24]	–	–	–
LAES	–	55%–80% [30]	–	25 + [30,39]	–
PCMs	–	75%–90% [33]	–	10–30 + [33]	–

Storage Technology	Technology Readiness Level	Storage Output Temperature (°C)
TCS	5 [24]	20–200 [37]
Molten salts	9 [38]	550 [29]
LAES	6 [39]	< 400 [39]
PCMs	4 [40]	−40–400 [33]

Table A.20 Environmental Impacts of Latent Thermal Energy Storage Systems

Environmental Impact	TCS	Molten Salts	LAES	PCMs
Land and water impact	Insignificant	Insignificant	Insignificant	Insignificant
Emissions produced during operation	None	None	Yes	None
Hazardous materials	Yes	Yes	None	Yes
Hazardous fumes	None	Yes	None	None
Short-term safety concerns	None	Minimal	None	None
Resource depletion	Insignificant	Insignificant	Insignificant	Insignificant
Geographic requirements	None	None	None	None

Table A.21 Compatible Applications for Latent Thermal Energy Storage Systems

Service	TCS	Molten Salts	LAES	PCMs
Energy arbitrage	Compatible	Compatible	Compatible	Compatible
Frequency regulation	Incompatible	Incompatible	Incompatible	Incompatible
Load following	Incompatible	Somewhat compatible	Compatible	Incompatible
Voltage support	Incompatible	Incompatible	Incompatible	Incompatible
Spinning reserves	Incompatible	Compatible	Compatible	Incompatible
Nonspinning and supp reserves	Somewhat compatible	Compatible	Compatible	Somewhat compatible
Black start	Incompatible	Incompatible	Compatible	Incompatible
VSR integration	Compatible	Compatible	Compatible	Compatible
Seasonal storage	Compatible	Incompatible	Compatible	Compatible
Process heat applications	Incompatible	Compatible	Somewhat compatible	Compatible

APPENDIX B: POLICY AND MARKET CONDITIONS FOR ENERGY STORAGE TECHNOLOGIES

In this section, federal and state regulations affecting energy storage technologies are detailed. In the selection methodology discussed in this chapter, a region was said to have positive policy conditions if the state regulations at the plant site were supportive of energy storage. Federal regulations are listed in Table B.1, and state regulations are listed in Table B.2.

Table B.1 Federal Regulations Affecting Energy Storage Technologies

Regulation	Description	Impact
FERC Order 719	Requires ISOs and RTOs to allow demand response resources to participate in energy and ancillary service markets. Also requires shorter intervals for price calculations, which better accounts for variability and favors energy storage [67]	This order is favorable for energy storage technologies that primarily act as demand response resources
FERC Order 745	Requires that electricity markets pay demand response resources at the market price for energy [67]	This order had the same effect as FERC Order 719

Table B.1 Federal Regulations Affecting Energy Storage Technologies—cont'd

Regulation	Description	Impact
FERC Order 755	The "pay-for-performance" order ensures that technologies providing regulation services are compensated according to the accuracy and speed of their response [67]	This order enables fast-responding energy storage technologies to receive more revenue for regulation than conventional generators
FERC Order 784	Expanded on the pay-for-performance order, FERC Order 755, by opening up ancillary services more broadly to energy storage participation [24]	This order further enhanced the profitability of energy storage and made the valuation of energy storage more transparent
FERC Order 890	Further opened up established energy markets to nongenerating resources such as demand response and energy storage [24]	Continued to create more markets for energy storage technologies to sell services
FERC Order 1000	This order requires public utility transmission providers to cooperate at a regional level. Neighboring regions must also coordinate to investigate all possible solutions to meet their requirements [68]	Regionally planned transmission and a clearer cost allocation process would open the market more to renewable energy developers, which could in turn drive the development of other emerging technologies
STORAGE Act of 2013	The Storage Technology for Renewable and Green Energy (STORAGE) Act of 2013 proposed a 30% ITC for businesses installing in-house energy storage systems and a 20% ITC for grid-connected installations [68]. This bill was not enacted	This policy would further enhance the economic viability of energy storage and encourage investors to pursue energy storage opportunities

Continued

Table B.1 Federal Regulations Affecting Energy Storage Technologies—cont'd

Regulation	Description	Impact
H. R. 5350, Energy Storage for Grid Resilience and Modernization Act	H. R. 5350 would establish a 30% ITC for both businesses and individuals interested in either producing or installing energy storage technologies [69]. This bill is currently before the House Ways and Means Committee	If passed, this federal policy would greatly enhance the economic feasibility of many energy storage technologies

Table B.2 State Policies and Initiatives Affecting Energy Storage Technologies

State	Policy or Initiative	Description
California	CPUC SGIP rules	The CPUC's Self-Generation Incentive Program provides financial incentives for consumer storage projects [26,68]. The SGIP was conceived in 2001
	AB 2514	Assembly Bill 2514 passed by the California state legislature tasked the CPUC with exploring energy storage initiatives. In response, the CPUC established a procurement target of 1.325 GW of storage by 2020 for all investor-owned utilities [67]. This bill was passed in 2010
Colorado	Innovative Clean Technology program	This program was founded by Colorado to provide funding for energy storage research and development [67]. This program was founded in 2009
	Section 123 resources	This initiative established by state law provides funding for emerging technologies without requiring that the technology be economically competitive [67]. This initiative was established by state law in 2001
Hawaii	–	Hawaii electric companies included energy storage in their 2013 Integrated Resource Plan (IRP), and Maui is considering energy storage as an option for addressing wind curtailment [65]

Table B.2 State Policies and Initiatives Affecting Energy Storage Technologies—cont'd

State	Policy or Initiative	Description
New Jersey	Clean Energy program	This program has $2.5 million of state funding for energy storage projects. However, they must be connected to a renewable energy source and ideally would primarily provide resiliency services. Although, the state's *Energy Master Plan* concluded that energy storage was not currently economically viable and recommended against pursuing energy storage as a resiliency solution [65]
	Critical infrastructure program	This program has an additional $500 million for updating existing infrastructure that could be used to build energy storage installations to defer upgrading the transmission infrastructure [65]
New Mexico	Energy Storage Task Force	This task force was formed to investigate investment options in energy storage for the state [65]
New York	NY-BEST	The New York Battery and Energy Storage Technology (NY-BEST) consortium provides funding for energy storage development and is supported by the New York State Energy Research and Development Authority (NYSERDA) [65]
	Green Bank initiative	The state has pledged almost $1 billion in financing for energy storage and other "green energy" projects [65]
	Energy Highway	This initiative was proposed by the state for the purpose of incentivizing the process of updating aging infrastructure [65]. In 2013, the implementation of this proposal began
Oregon	–	Portland General Electric included energy storage in their 2013 RFP (request for proposals), opening the door for investment in energy storage installations [65]
Texas	SB 943	This bill required energy storage installations to be registered as generation assets when used to sell energy or ancillary services, limiting the benefits that energy storage can provide in the state [65]. This bill became law in 2011

Continued

Table B.2 State Policies and Initiatives Affecting Energy Storage Technologies—cont'd

State	Policy or Initiative	Description
	Texas Docket 39917	TD 39917 required energy storage charging and discharging to be considered wholesale energy transactions. This change improved the economics of energy storage and eliminated several market distortions in the location and operation of resources [67]. This initiative was issued in 2012
Washington	–	The Washington Utilities and Transportation Commission requested that utilities in the state include energy storage when considering resource options for their next IRP [65]
Other states	–	Connecticut, Maryland, and Maine are also evaluating energy storage and microgrid development as options to improve grid resiliency and enable smart grid technologies [65]

As described earlier in the chapter, regional market conditions were assessed by calculating the variability of electricity generation in each region. This was done by calculating the percentage of electricity generation from variable renewable resources in each NERC subregion. The results are displayed in Table B.3.

Table B.3 Market Variability in Certain Regions of the United States [70]

Region	Percentage of Wind and Solar Electricity Generation by Region
New England	3.58
Middle Atlantic	2.36
East North Central	3.90
West North Central	15.77
South Atlantic	0.63
East South Central	0.06
West South Central	8.66
Mountain	7.28
Pacific contiguous	12.63
Pacific noncontiguous	9.04

APPENDIX C: ENERGY STORAGE COST COMPARISONS

A large amount of data is contained in the tables shown previously in Appendix A, so these metrics are displayed graphically in the following figures for ease of comparison.

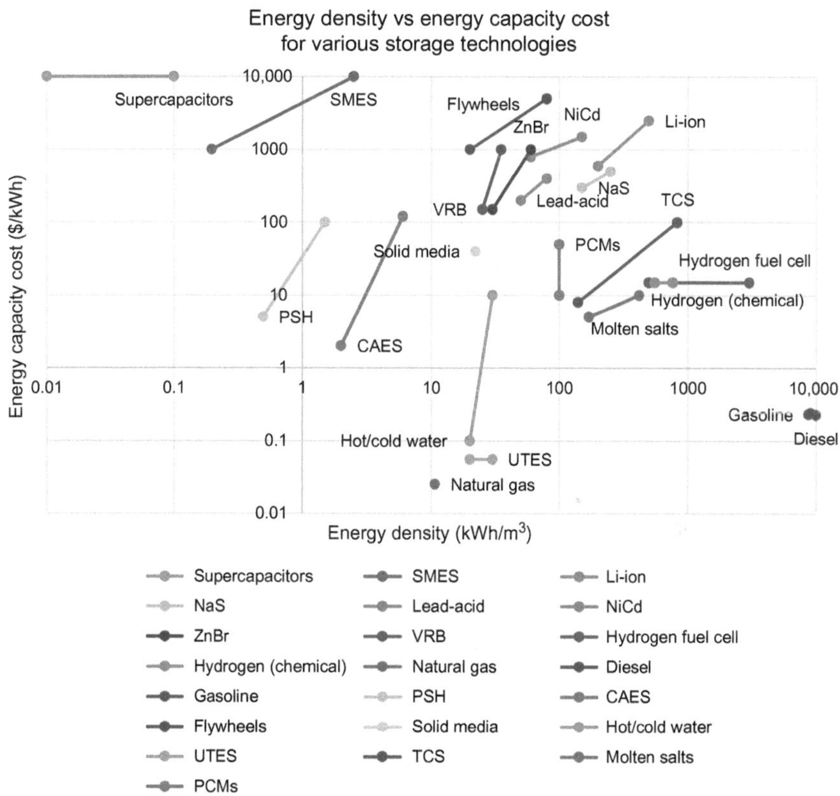

Fig. C.1 Comparing energy density and cost of storage for various technologies.

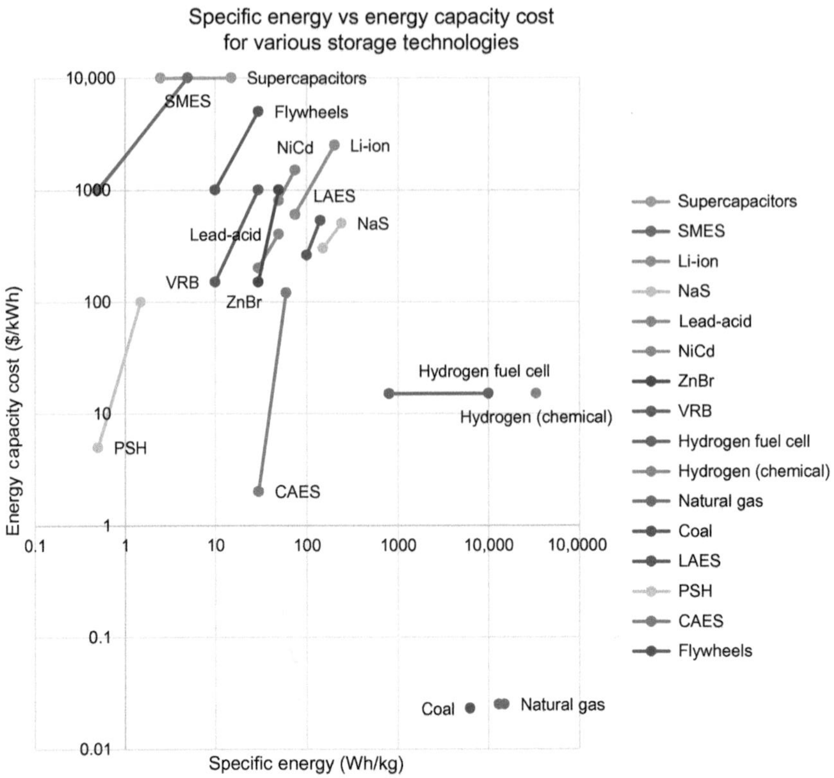

Fig. C.2 Comparing specific energy and cost of storage for various technologies.

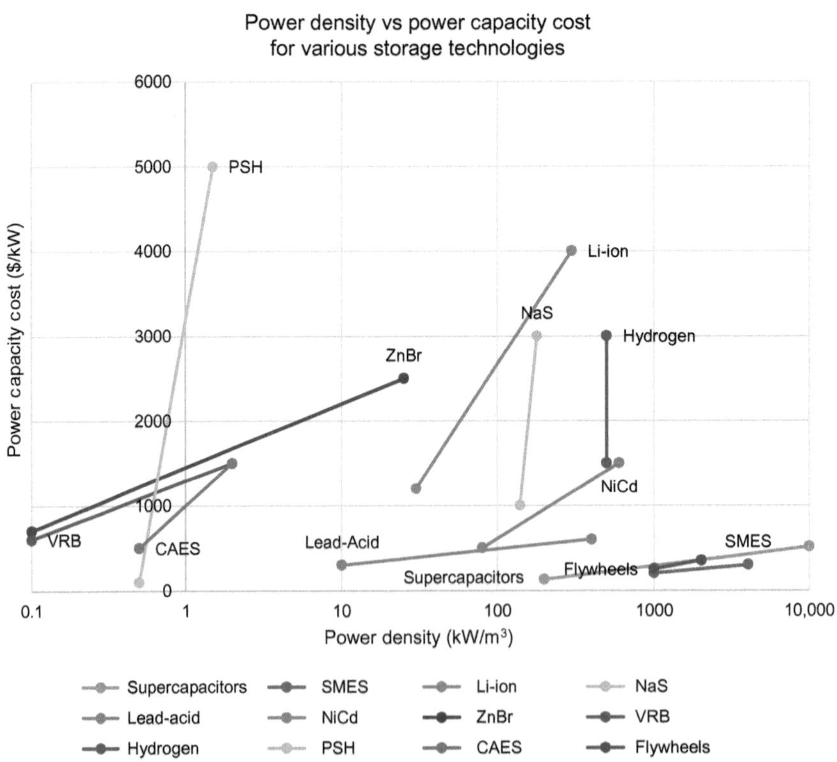

Fig. C.3 Comparing power density and cost of power for various technologies.

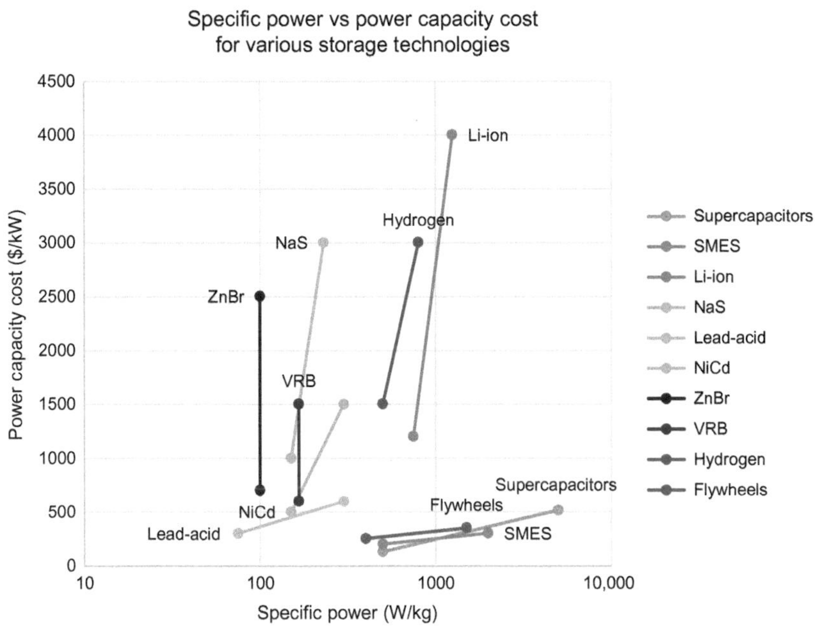

Fig. C.4 Comparing specific power and cost of power for various technologies.

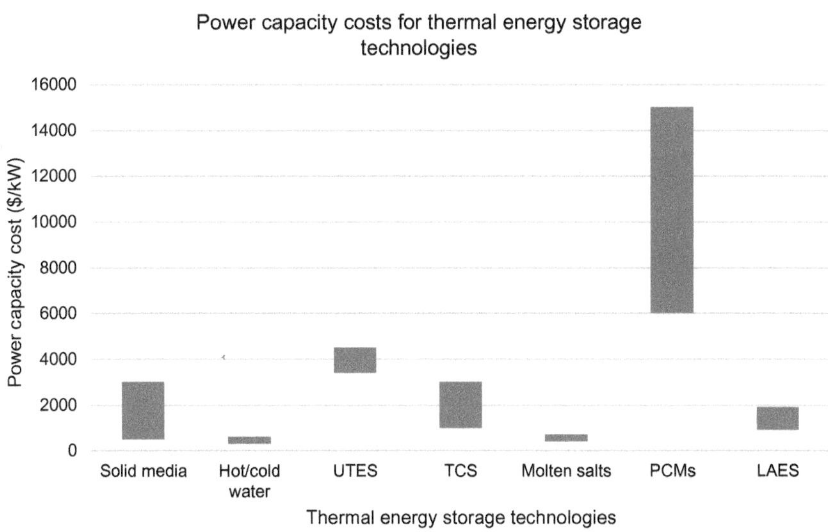

Fig. C.5 Comparing the cost of power for thermal energy storage technologies.

APPENDIX D: DETAILED SELECTION METHODOLOGY

A selection methodology for comparing energy storage technologies for integration with advanced NPPs was outlined generally in this chapter. More detailed equations for calculating "scores" for each of the selection criteria are included here.

Technical maturity: Starting from each technology's TRL score, a normalization function was used to calculate a score on a scale from 1 to 10. The technical maturity score is calculated as

$$T = \overset{[0,\,10]}{N}\left(\overset{[0,\,9]}{\mathrm{TRL}}\right) \cdot \left. \left(\frac{1 + \overset{[1,\,5]}{W} - 1}{5} \right) \right|_{\max=10} \tag{D.1}$$

$$N(x) = \frac{x}{\max(x)} \times 10 \tag{D.2}$$

where T represents the technical maturity score, $N(x)$ represents the normalization function, TRL represents the technology readiness level, and W represents a weighting factor. This weighting factor is used to increase a storage technology's technical maturity by approximately one point for every 5 years between when the score is calculated and when the technology will be installed, which accounts for any future development.

Economic feasibility: The economic feasibility score is calculated by comparing the installed cost of an energy storage technology to an NPP's available budget. If the cost of the technology is greater than the NPP's budget, then the economic feasibility score will be zero. The least expensive energy storage option is given a score of 10, and the rest of the technologies are assigned a score between 0 and 10. The economic feasibility score is calculated as

$$\begin{aligned} IC &= \max\left(C_E \cdot E_{req},\, C_P \cdot P_{req}\right), \\ EF &= \begin{cases} 0, & \text{if } B - IC < 0 \\ \overset{[0,\,10]}{N}\left(\dfrac{B - IC}{B}\right), & \text{if } B - IC \geq 0 \end{cases} \end{aligned} \tag{D.3}$$

where P_{req} and E_{req} represent the NPP's power and energy requirements, C_P and C_E represent the energy storage system's power and energy capacity costs, IC represents the total installed cost for a storage technology, EF

represents a storage technology's economic feasibility score, and B represents the NPP's budget.

Environmental impact: First, some environmental impact parameters must be weighted. For example, if the energy storage technology will be installed at a safe distance from sensitive populations, the "use of hazardous materials" parameter should be weighted as less significant. This weighting is done using the factors shown in Table D.1.

Table D.1 Weighting Factors for the "Use of Hazardous Materials" Parameter

Proximity to Sensitive Populations (ft)	Weighting Factor
500+	0.2
300–400	0.4
200–300	0.6
100–200	0.8
0–100	1

The parameters in Table 5.3 are then normalized to a scale from 0 to 1 (using Eq. D.2) and averaged together. Once the parameter set has been averaged, the resulting value is set to a scale from 0 to 10 and subtracted from 10 so that a higher score represents a storage technology with a less significant environmental impact. This is the reverse of the scale used for the parameters listed in Table 5.3. The scale was inverted so that a higher score in this category would correlate to a more appropriate energy storage system, as is the case with the *technical maturity* and *economic feasibility* scores. A technology's overall environmental impact score is calculated as

$$EI = 10 - \overset{[0,\,10]}{N}\left(\sum\left[\frac{\overset{[0,\,1]}{W}\cdot\overset{[0,\,1]}{\varepsilon}}{6}\right]\right) \tag{D.4}$$

where EI represents the environmental impact score, N represents the normalization function, W represents a weighting factor, and ε represents the environmental impact parameters listed in Table 5.3.

ACKNOWLEDGMENTS

This chapter is a product of the efforts of individuals at both The University of Texas at Austin and Idaho National Laboratory. The authors of this chapter would like to acknowledge the US Department of Energy and, specifically, the Idaho National Laboratory, for funding this research. Battelle Energy Alliance, LLC operates Idaho National Laboratory under contract no. DE-AC07-05ID14517 with the US Department of Energy.

REFERENCES

[1] W. Cole, et al., 2016 Standard Scenarios Report: A U.S. Electricity Sector Outlook, National Renewable Energy Laboratory, Golden, CO. 2016.

[2] P. Maloney, How market forces are pushing utilities to operate nuclear plants more flexibly, Utility Dive, [Online]. Available from: http://www.utilitydive.com/news/how-market-forces-are-pushing-utilities-to-operate-nuclear-plants-more-flex/427496/. 2016.

[3] Nuclear Energy Institute, Nuclear plant shutdowns reveal market problems, [Online]. Available from: https://www.nei.org/News-Media/News/News-Archives/Nuclear-Plant-Shutdowns-Reveal-Market-Problems, 2014.

[4] N. Kumar, P. Besuner, S. Lefton, D. Agan, D. Hilleman, Power plant cycling costs, National Renewable Energy Laboratory, Golden, CO, 2012.

[5] NEA, Nuclear Energy and Renewables: System Effects in Low-Carbon Electricity Systems, Nuclear Energy Agency, Paris, France,, 2012.

[6] U.S. Energy Information Administration, What is U.S. electricity generation by energy sources?, [Online]. Available from: https://www.eia.gov/tools/faqs/faq.php?id=427&t=3, 2018.

[7] Nuclear Energy Institute, Environment: emissions prevented, [Online]. Available from: https://www.nei.org/Knowledge-Center/Nuclear-Statistics/Environment-Emissions-Prevented, 2016.

[8] B.K. Sovacool, Valuing the greenhouse gas emissions from nuclear power: a critical survey, Energy Policy 36 (8) (2008) 2940–2953.

[9] J.R. Lovering, A. Yip, T. Nordhaus, Historical construction costs of global nuclear power reactors, Energy Policy 91 (2016) 371–382.

[10] B. Mann, Unable to Compete on Price, Nuclear Power on the Decline in the U.S., *National Public Radio*Washington, DC, 2016. [Online]. Available from: http://www.npr.org/2016/04/07/473379564/unable-to-compete-on-price-nuclear-power-on-the-decline-in-the-u-s.

[11] U.S. Energy Information Administration, Electricity data browser, [Online]. Available from: https://www.eia.gov/electricity/data/browser/, 2018.

[12] World Nuclear Association, Nuclear power in the USA, [Online]. Available from: http://www.world-nuclear.org/information-library/country-profiles/countries-t-z/usa-nuclear-power.aspx, 2018.

[13] IAEA, International Status and Prospects for Nuclear Power 2017, International Atomic Energy Agency, Vienna, Austria, 2017.

[14] The Aspen Institute, Reducing carbon emissions from electricity generation, in: Energy: Old Challenges, New Opportunities, The Aspen Institute, Washington, DC, 2009, pp. 27–38.

[15] M.Z. Jacobson, et al., 100% clean and renewable wind, water, and sunlight (WWS) all-sector energy roadmaps for 139 countries of the world, Joule 1 (2017) 108–121.

[16] P. Luckow, B. Fagan, S. Fields, M. Whited, Technical and Institutional Barriers to the Expansion of Wind and Solar Energy, Synapse Energy Economics, Cambridge, MA, 2015.

[17] J. Eyer, G. Corey, Energy Storage for the Electricity Grid: Benefits and Market Potential Assessment Guide, Sandia National Laboratories, Albuquerque, NM, 2010.

[18] J.D. Jenkins, et al., The benefits of nuclear flexibility in power system operations with renewable energy, Appl. Energy 222 (2018) 872–884.

[19] World Nuclear Association, Advanced nuclear power reactors, [Online]. Available from: http://www.world-nuclear.org/information-library/nuclear-fuel-cycle/nuclear-power-reactors/advanced-nuclear-power-reactors.aspx, 2017.

[20] IAEA, Status Report 81—Advanced Passive PWR (AP 1000), International Atomic Energy Agency, Vienna, Austria, 2011.
[21] F.J. de Sisternes, J.D. Jenkins, A. Botterud, The value of energy storage in decarbonizing the electricity sector, Appl. Energy 175 (2016) 368–379.
[22] W.A. Braff, J.M. Mueller, J.E. Trancik, Value of storage technologies for wind and solar energy, Nat. Clim. Chang. 6 (October) (2016).
[23] C. Forsberg, et al., Light Water Reactor Heat Storage for Peak Power and Increased Revenue: Focused Workshop on Near-Term Options, Massachusetts Institute of Technology, Cambridge, MA, 2017.
[24] M.C. Lott, S.-I. Kim, Technology Roadmap: Energy Storage, International Energy Agency, Paris, France, 2014.
[25] U.S. Department of Energy, Hydropower Vision: A New Chapter for America's 1st Renewable Energy Source, U.S. Department of Energy, Washington, DC, 2016.
[26] A.A. Akhil, et al., DOE/EPRI Electricity Storage Handbook in Collaboration With NRECA, Sandia National Laboratories, Albuquerque, NM, 2015.
[27] I. Gyuk, et al., Grid Energy Storage, U.S. Department of Energy, Washington, DC, 2013.
[28] M. Beaudin, H. Zareipour, A. Schellenberglabe, W. Rosehart, Energy storage for mitigating the variability of renewable electricity sources: an updated review 10. Energy Sustain. Dev. 14 (4) (2010) 302–314, https://doi.org/10.1016/j.esd.2010.09.007.
[29] J. Intrator, et al., 2020 Strategic Analysis of Energy Storage in California, California Energy Commission, Sacramento, CA, 2011.
[30] X. Luo, J. Wang, M. Dooner, J. Clarke, Overview of current development in electrical energy storage technologies and the application potential in power system operation, Appl. Energy 137 (2015) 511–536.
[31] D. Ferrero, A. Lanzini, M. Santarelli, P. Leone, A comparative assessment on hydrogen production from low- and high-temperature electrolysis, Int. J. Hydrog. Energy 38 (9) (2013) 3523–3536.
[32] A. Körner, Technology Roadmap: Hydrogen and Fuel Cells, International Energy Agency, Paris, France, 2015.
[33] IRENA and IEA-ETSAP, Thermal energy storage – technology brief, IRENA and IEA-ETSAP, Abu Dhabi, UAE, and Paris, France, 2013.
[34] W.-D. Steinmann, Thermal Energy Storage Systems for Concentrating Solar Power (CSP) Plants, in: Concentrating Solar Power Technology: Principles, Developments and Applications, Woodhead Publishing Series in Energy, Cambridge, UK, 2012, pp. 362–394.
[35] D. Laing, C. Bahl, T. Bauer, M. Fiss, N. Breidenbach, M. Hempel, High-temperature solid-media thermal energy storage for solar thermal power plants, Proc. IEEE 100 (2) (2011) 516–524.
[36] C.W. Forsberg, Gigawatt-year geothermal energy storage coupled to nuclear reactors and large concentrated solar thermal systems, in: PROCEEDINGS, Thirty-Seventh Workshop on Geothermal Reservoir Engineering Stanford, CA, 2012.
[37] A.H. Abedin, M.A. Rosen, A critical review of thermochemical energy storage systems, *Open Renew.* Energy J. 4 (2011) 42–46.
[38] C. Philibert, Technology Roadmap: Concentrating Solar Power, International Energy Agency, Paris, France, 2010.
[39] G. Brett, M. Barnett, The application of liquid air energy storage for large scale long duration solutions to grid balancing, EPJ Web Conf. 79 (2014) 03002.
[40] A. Sharma, V.V. Tyagi, C.R. Chen, D. Buddhi, Review on thermal energy storage with phase change materials and applications, Renew. Sust. Energ. Rev. 13 (2) (2009) 318–345.

[41] U.S. Department of Energy, Technology Readiness Assessment Guide, U.S. Department of Energy, Washington, DC, 2011.

[42] Lazard, Lazard's Levelized Cost of Storage Analysis—Version 1.0, Lazard, Hamilton, Bermuda, 2015.

[43] S.S. Raza, I. Janajreh, C. Ghenai, Sustainability index approach as a selection criteria for energy storage system of an intermittent renewable energy source, Appl. Energy 136 (2014) 909–920.

[44] Sandia Corporation, DOE Global Energy Storage Database, [Online]. Available from: http://www.energystorageexchange.org/, 2018.

[45] P. Denholm, E. Ela, B. Kirby, M. Milligan, The role of energy storage with renewable electricity generation, National Renewable Energy Laboratory, Golden, CO, 2010.

[46] J. Lubek, S. Wakeford, The Future of Hydroelectricity in France: What Does the New Energy Transition Law Mean for the Long-Delayed Renewal of Concessions? NERA Economic Consulting, White Plains, NY, 2015.

[47] D.V. Syranov, V.N. Kovalnogov, A.N. Zolotov, Modeling, research and optimization of heat losses during transport in energy systems, in: 2nd International Conference on Industrial Engineering, Applications and Manufacturing, Chelyabinsk, Russia, 2016.

[48] Georgia Power, Fact Sheet: Vogtle Units 3 & 4, Southern Company, Atlanta, GA, 2018.

[49] Southern Company, Southern Company subsidiary Georgia Power files recommendation to complete construction of Vogtle nuclear expansion, *News Center Stories* (2017) [Online]. Available from: https://www.southerncompany.com/newsroom/2017/aug-2017/georgia-power-vogtle-recommendation.html.

[50] R.B. Schainker, A. Rao, Compressed Air Energy Storage Scoping Study for California, California Energy Commission, Sacramento, CA, 2008.

[51] B. Hadjerioua, W. Yaxing, S.-C. Kao, An Assessment of Energy Potential at Non-Powered Dams in the United States, U.S. Department of Energy, Washington, DC, 2012.

[52] U.S. Geological Survey, "Aquifers: map of the principal aquifers of the United States. *USGS Groundwater Information*, 2017. [Online]. Available from: https://water.usgs.gov/ogw/aquifer/map.html.

[53] National Renewable Energy Laboratory, Geothermal Resource of the United States, *Geothermal Maps* (2009) [Online]. Available from: https://www.nrel.gov/gis/geothermal.html.

[54] D.C. Culver, H.H. Hobbs, M.C. Christman, L.L. Master, Distribution map of caves and cave animals in the United States, J. Cave Karst Stud. 61 (3) (1999) 139–140.

[55] Southern Company, Open Access Transmission Tariff of Alabama Power Company, Georgia Power Company, Gulf Power Company, and Mississippi Power Company (Southern Company), Southern Company, Atlanta, GA, 2018.

[56] J. Coleman, et al., An evaluation of energy storage options for nuclear power, Idaho National Laboratory, Idaho Falls, ID, 2017.

[57] D. Rastler, Electric energy storage technology options: a white paper primer on applications, costs, and benefits, Electricity Power Research Institute, Palo Alto, CA, 2010.

[58] S.M. Schoenung, Energy storage systems cost update, (2011).

[59] Z.-S. Wu, K. Parvez, X. Feng, K. Müllen, Graphene-based in-plane micro-supercapacitors with high power and energy densities, Nat. Commun. 4 (2013).

[60] D. Pech, et al., Ultrahigh-power micrometre-sized supercapacitors based on onion-like carbon, Nat. Nanotechnol. 5 (9) (2010) 651–654.

[61] S. Mccluer, J.-F. Christin, Comparing data center batteries, flywheels, and ultracapacitors, APC by Schneider Electric, West Kingston, RI, 2011.

[62] W.G. Manuel, Turlock Irrigation District: Energy storage study 2014, California Energy Commission, Sacramento, CA, 2014.

[63] D.B. Gray, Tesla switches on World's biggest lithium ion battery, Sci. Am. (2017) [Online]. Available from: https://www.scientificamerican.com/article/tesla-switches-on-world-rsquo-s-biggest-lithium-ion-battery/.
[64] K. Gaine, A. Duffy, A life cycle cost analysis of large-scale thermal energy storage technologies for buildings using combined heat and power, in: Zero Emission Buildings, Conference Proceedings, Trondheim, Norway, 2010.
[65] W.D. Steinmann, M. Eck, Buffer storage for direct steam generation, Sol. Energy 80 (10) (2006) 1277–1282.
[66] L.F. Cabeza, I. Martorell, L. Miró, A.I. Fernández, C. Barreneche, Introduction to thermal energy storage (TES) systems, in: L.F. Cabeza (Ed.), Advances in Thermal Energy Storage Systems, Elsevier Ltd., Amsterdam, 2015, pp. 1–28
[67] D. Bhatnagar, A. Currier, J. Hernandez, O. Ma, Market and policy barriers to energy storage deployment: a study for the energy storage systems program, Sandia National Laboratories, Albuquerque, NM, 2013.
[68] D. Bhatnagar, Federal and state energy storage policies and efforts, New Mexico Energy Storage Task Force, Albuquerque, NM, 2013.
[69] M.M. Honda, H.R. 5350—Energy Storage Act of 2016, [Online]. Available from: https://www.congress.gov/bill/114th-congress/house-bill/5350/text, 2016.
[70] U.S. Energy Information Administration, Electricity Power Monthly With Data for 2015, U.S. Energy Information Administration, Washington, DC, 2016.

CHAPTER SIX

Chemical Energy Storage

Shripad T. Revankar
School of Nuclear Engineering, Purdue University, West Lafayette, IN, United States

6.1 INTRODUCTION

6.1.1 Energy Storage Systems and Need

The majority of current energy need in the world is met by the fossil fuel due to its availability, economics, and the infrastructure to use it in major four sectors: industrial, transportation, residential, and commercial. The other two primary energy sources, renewable and nuclear, which have no greenhouse gas emissions contribute remaining share of the energy needs. The continued use of fossil fuel has created serious concerns regarding environment, limited availability and geographic constraints, and energy security concerns. This has spurred interest in generating energy from renewable sources specifically solar and wind energy that are most abundant and potentially readily available. However, renewable sources are intermittent and are not reliable sources of power. In addition, the renewable resources are localized and may be far-off from city centers and hence have to rely on electric grid lines. The variability of renewables presents a great challenge in energy generation and load balance maintenance to ensure power network stability and reliability. Common to most renewable energy sources is that their location is resource-dependent and transmissions systems need to be developed to accommodate this constraint. The baseload power plants such as fossil or nuclear need to compensate for the variability. Thus, there are three main characteristics that need to be taken into consideration in the process of integrating increased renewable energy-based power supply: (i) variability and predictability (uncertainty); (ii) location, relevant for the network design; and (iii) capacity factor, capacity credit, and power plant characteristics, which are important when comparing the renewable energy sources with thermal fossil or nuclear plants.

Energy storage technologies can offset the intermittency problem of renewable energy sources by storing the generated intermittent energy

Storage and Hybridization of Nuclear Energy
https://doi.org/10.1016/B978-0-12-813975-2.00006-5

and then making it accessible upon demand. In addition, the energy storage system can also provide many grid services such as (i) frequency regulation and load following (aggregated term often used is balancing services), (ii) - cold-start services, (iii) contingency reserves, and (iv) energy services that shift generation from peak to off-peak periods. Energy storage is a key enabler of the smart grid or future grid, which is expected to integrate a significant amount of renewable energy generation, transmission, and distribution. The energy storage technologies also have the potential to transform the transportation system where energy storage devices could replace the power train systems of current transportation technologies from a chemical fuel-based power train to an electricity-based power train. The storage technologies can bridge temporal and geographic gaps between energy supply and demand when coupled with other energy infrastructure components. In a broader sense, energy storage is a system integration technology that facilitates improved management of energy supply and demand. A single unit of energy storage infrastructure can provide multiple valuable energy and power services as heat and electricity. Energy storage technologies are valuable components in most energy systems and could be an important tool in achieving a low-carbon future. These technologies allow for the decoupling of energy supply and demand, in essence providing a valuable resource to system operators. Energy storage technologies can be implemented on large and small scales in distributed and centralized manners throughout the energy system. Also, the time scale in energy storage devices varies from seconds to years. The electric vehicle is a best example of how energy storage technologies can support sustainable transportation system. Electronic devices such as smartphones and laptops, which have widespread use around the globe, are also heavily reliant on energy storage technologies.

In energy storage technologies, energy in the form of either chemical, thermal, electric, or kinetic is absorbed and is stored for a period of time before releasing it to supply energy or power services. The energy can be transformed to many different forms for storage:

(1) As gravitational potential energy using mechanical pumps with water reservoirs.
(2) As compressed air using air compressors.
(3) As kinetic energy in flywheels.
(4) As electrochemical energy in batteries, chemical capacitors, and flow batteries.
(5) As chemical energy in a hydrogen, hydrocarbons, or other chemicals.
(6) As magnetic field in inductors.

(7) As electric field in capacitors.
(8) As thermal energy either as sensible or latent heat in materials.
Fig. 6.1 shows the classification of the energy storage technologies in the form of energy stored, mechanical, chemical, electric, and thermal energy storage systems. Among these, chemical energy storage (CES) is a more versatile energy storage method, and it covers electrochemical secondary batteries; flow batteries; and chemical, electrochemical, or thermochemical processes based on various fuels such as hydrogen, synthetic natural gas (SNG), methane, hydrocarbons, and other chemicals products. In addition to chemical batteries, it includes chemical capacitors as well. Two well-known storage technologies of the existing energy system are heat storage in combined heat and power (CHP) in cogeneration systems and water reservoirs in hydropower systems. A CHP plant meets demand ofboth heat and electricity by including heat storage, reducing the operating constraints on the power side of the system, and allowing it to be dispatched as required by the electric needs. Heat storage capacity helps smooth the power regulation. Hydropower systems with reservoirs can supply storage services on time scales ranging from seconds to months or even years. Moreover, this large-scale storage can meet demand peaks, balance variations in other forms of generation, or stabilize output from the hydroplant as rainfall varies between seasons or consecutive years. While some technologies are mature or near maturity, most are still in the early stages of development specifically for a large scale and will require additional attention before their potential can be fully realized.

Fig. 6.2 shows the comparison of rated power and rated energy capacity of various energy storage technologies and their range of discharge times. Energy storage technologies and systems are diverse. These storage methods can be classified by the nominal discharge time at rated power: (i) discharge

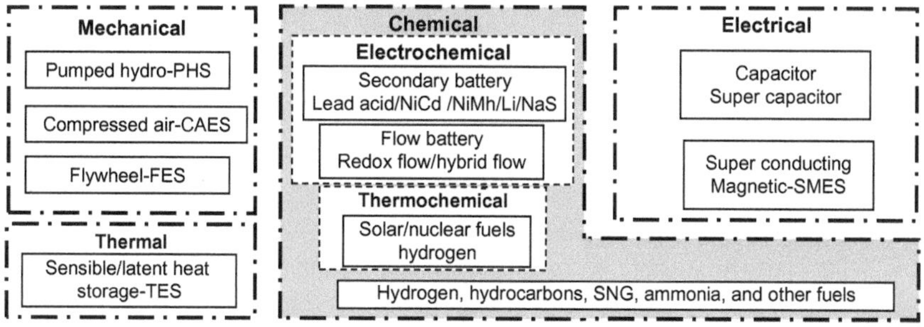

Fig. 6.1 Energy storage technologies by the form of energy stored.

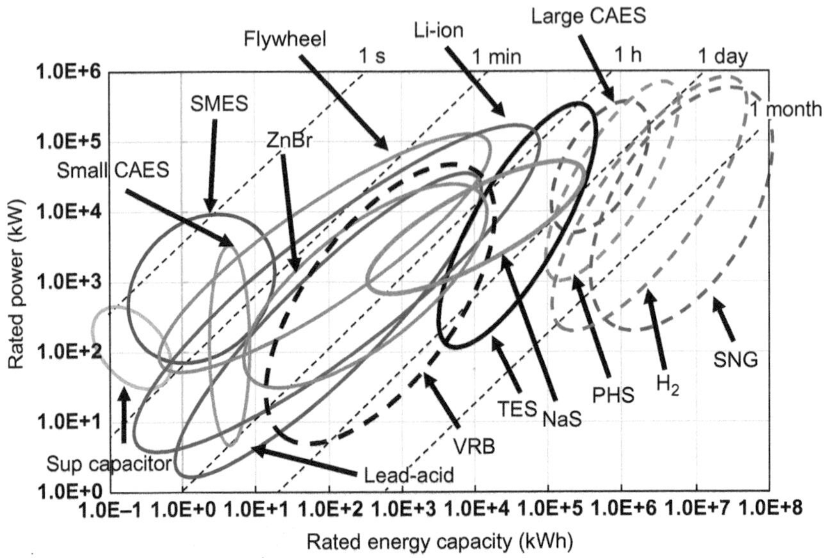

Fig. 6.2 Rated energy capacity and rated power for various energy storage system with discharge time durations. *CAES*, compressed air energy storage; *H_2*, hydrogen; *Li-ion*, lithium-ion batteries; *Na-S*, sodium-sulfur batteries; *PHS*, pumped hydro storage; *SMES*, superconducting magnetic energy storage; *SNG*, synthetic natural gas, *TES*, thermal energy storage; *VRB*, vanadium redox flow batteries; *ZnBr*, zinc-bromide batteries.

time <1 h such as flywheel, supercapacitor, and superconducting magnetic energy storage; (ii) discharge time up to around 10 h: aboveground small-scale compressed air and various batteries including lead-acid, lithium-ion, nickel-cadmium, and zinc-bromide; (ii) discharge time longer than 10 h: pumped hydro, underground large-scale compressed air, liquid air energy storage, vanadium redox flow batteries, thermal energy storage, fuel cell, hydrogen, SNG, and hydrocarbons. The energy storage capacities of the systems range from fraction of kilowatt-hour with supercapacitors to hundreds gigawatt-hour storage with power capacity to gigawatts. The chemical energy storage technologies such as hydrogen, SNG, and other hydrocarbons provide large energy storage capacity. Table 6.1 shows the key energy storage features of various technologies.

There are technological and economic challenges for transition to a non-fossil energy system with a high share of renewable energy sources such as solar and wind. The capital investment costs of energy storage are considerable with current storage technologies for meeting consumption and production. However, this condition is likely to change in the future as storage costs decrease and the grid stability become increasingly important.

Table 6.1 Features of Energy Storage Technologies

Storage Technology	Energy Density (Wh/kg)	Energy Density (Wh/L)	Power Density (W/L)	Typical Discharge Time	Efficiency (%)	Typical Applications
Pumped hydro	0.2–2	0.2–2	0.1–0.2	Hours	70–80	Time shifting, power quality, emergency supply
Compressed air	–	2–6	0.2–0.6	Hours	41–75	Time shifting
Flywheel	5–30	20–80	5000	Seconds	80–90	Power quality
Lead-acid	30–45	50–80	90–700	Hours	75–90	Off-grid, emergency supply, time shifting, power quality
NiCd vented	15–40	15–80	75–700	Hours	60–80	Off-grid, emergency supply, time shifting, power quality
Sealed	30–45	80–110	(vented)	Hours	60–70	Off-grid, emergency supply, time shifting, power quality
NiMH sealed	40–80	80–200	500–3000	Hours	65–75	Electric vehicle
Li-ion	60–200	200–400	1300–10,000	Hours	85–98	Power quality, network efficiency, off-grid, time shifting, electric vehicle
Zinc air	130–200	130–200	50–100	Hours	50–70	Off-grid, electric vehicle
Sodium-sulfur	100–250	150–300	120–160	Hours	70–85	Time shifting, network efficiency, off-grid
NaNiCl	100–200	150–200	250–270	Hours	80–90	Time shifting, electric vehicles
Vanadium redox flow battery	15–50	15–50	0.5–2	Hours	60–75	Time shifting, network efficiency, off-grid
Hybrid flow battery	75–85	75–85	1–25	Hours	65–75	Time shifting, network efficiency, off-grid
Hydrogen central	33,330	33,330	0.2–2	Hours to weeks	34–44	Time shifting
Decentral			0.2–20			
Synthetic natural gas	10,000	1800 (200 bar)	0.2–2	Hours to weeks	30–38	Time shifting
Double-layer capacitor	1–15	10–20	40,000–120,000	Seconds	85–98	Power quality, effective connection
Superconducting magnetic energy storage	–	6	2600	Seconds	75–80	Time shifting, power quality

In addition, storage also creates the extensive benefits to the electricity system as a whole and can potentially play an important role in providing efficient, clean, and better access to energy and associated services for the global community.

6.1.2 Role of Chemical Energy Storage

As seen from Fig. 6.2, chemical energy storage technologies are mainly constituted by batteries (secondary and flow batteries) and renewable generated chemicals (hydrogen, fuel cell, SNG, and hydrocarbons). Batteries as electrochemical energy storage bring great promise in a range of small-scale to large-scale applications. The applications span from portable consumer electronics such as mobile phones, game pads, and tablet computers via mobile applications to electric vehicles, forklift trucks, boats, and emergency power in power plants at large-scale grid-based applications for stabilization and load balancing of intermittent, renewable sources like solar and wind energy. The battery technology can also be applied to aviation, distant transport of goods, and seasonal energy storage with further developments. Among batteries, the Li-ion chemistry can be used for low- and moderate-energy-density applications, such as in typical portable applications. The lithium-based batteries have potential large market share of batteries for transportation and mobility. As the usage of portable devices powered by battery increases and develops into new applications, a demand for new and specialized battery will be inevitable. For the sustainable energy infrastructure of the future, the greatest prospect for electrochemical storage is mostly on the stabilization of frequency and voltage in dealing with hourly and daily fluctuations. Battery technologies can be competitive for transport and stationary applications with substantial improvements in existing battery technologies and the development of novel high-performance batteries.

The chemicals have much greater energy density than the energy density of current battery technologies and have large discharge times since they can be stored for any time period. Hence, the applications of chemicals are varied, and specifically, they can be used for large-scale systems with applications for large duration. They can be used as raw materials for the chemical industry, for direct electricity production, and for the transportation sector as replacement fuel instead of fossil fuel. Synthetic fuels produced from sustainable energy can compliment or supplement batteries in the transportation sector. Examples are, hydrogen produced from solar energy, the forming of ammonia with nitrogen and methane, or methanol produced by electrochemical fixation of carbon dioxide. Thus, there will be substantial demand for chemical energy storage for the future renewable-resource-based energy sector.

In electric energy sector, excess electricity can be used to produce hydrogen with electrolysis to stabilize electric power to cope with demand changes. Hydrogen is an ideal molecule either to store itself as energy storage chemical or to process other storage molecules such as liquid hydrocarbons. Gasified biomass and carbon-containing waste fractions are other resources of renewable energy that can be used in the stabilization of fluctuating electricity production if produced in large capacity. Gasification is a well-known technology where the reaction of carbonaceous raw materials with steam at high temperature is carried out to produce SNG (mainly CO, CO_2, and H_2). SNG can be used to leverage the demand in electricity. During the period when electricity demands are met by renewables such as the sun and wind, then the SNG can be processed using catalytic processes into various fuels and chemicals such as methane, methanol, and synthetic gasoline and diesel. When there is more demand for electricity, the SNG can be directly used to produce electricity though combustion.

The syngas immediately produced in gasification has low hydrogen content than required for use as SNG; hence, it should be boosted with hydrogen, which can be produced from electrolysis of water. Similarly, many types of biomass have low energy density that can be upgraded by treatment with hydrogen, which in turn can be produced from water electrolysis. These technologies rely on catalysis and electrolysis that will play pivotal roles in storage routes. The chemical energy storage in the form of gaseous hydrogen or methane facilitate synthesis of SNG and hydrogen produced from electrolysis to liquid fuels such as dimethyl ether, methanol, and other liquid hydrocarbons to supply fuels to sectors such as aviation and heavy road transport. The production of liquid hydrocarbons fuels from electrolysis generated hydrogen and biomass would help reduce the dependence on fossil oil and to expand the use of surplus electric power into the aviation and heavy road transport segment.

6.2 LEAD-ACID AND LEAD-CARBON BATTERIES

6.2.1 Lead-Acid Batteries

Batteries are well-established energy storage methods currently existing in the marketplace. The batteries store electrochemical energy, which can be charged with external electric power and discharged to power as needed.

A unit cell of a battery consists of two electrodes with positive and negative polarity submersed in an electrolyte and a path for electrons motion, usually conducting wire connected to each electrode. A battery consists of multiple unit cells that are coupled in parallel or in series or both, to attain the preferred power rating and voltage level. In a unit cell of a battery, there is transfer of electrons in two physically separated chemical reactions, namely, the oxidation reaction and a reduction reaction. At oxidation site, electrons are given away and are received at reduction site. The battery current is generated when electrons transfers from one electrode to the other. When load is inserted between these two reaction sites, the electron transfer generates electric power. The electrolyte facilitates exchange of anions and cations (charged ions) between the oxidation and reduction reactions. The voltage of the battery is the difference between the electrochemical potential between the two electrodes, which correspond to the electrochemical reactions occurring at each electrode-electrolyte interface. For a discharging battery, the electrode where the oxidation reaction takes place is defined as the anode and has a positive voltage, and the electrode where the reduction reaction takes place is called the cathode and has a negative voltage. During discharging, the oxidation reaction at the anode generates electrons, and reduction reaction at the cathode uses these electrons, and therefore, during discharging, electrons flow from the anode to the cathode. The redox reactions that comprise a particular battery system define many fundamental parameters about the battery system. Other key battery properties, including battery capacity, charging/discharging performance, and other practical considerations are also influenced by the physical configuration of the battery, for example, the amount of electrolyte material in the battery or the geometry of the electrodes such as surface area.

Lead-acid batteries are the oldest and world's most widely used rechargeable electrochemical devices and have been commercially deployed since about 1890. They are based on the reversible electrochemical conversion of lead to lead sulfate (anode) and quadrivalent lead oxide to lead sulfate (cathode) in a concentrated sulfuric acid electrolyte. The electrodes are generally formed by a detailed process to give a thick coating (~1 mm) of lead compounds and additives on current collectors (grids) of lead, alloys, or carbons. Fig. 6.3 shows a single cell at two stages of its life cycle. Its electrodes are lead plates constructed in the form of grids to maximize their surface area. The electrodes are immersed in an electrolyte of aqueous sulfuric acid. As shown in Fig. 6.3A, the cell is fully charged and acting as a galvanic cell to supply electric current to an external load. During this process, the

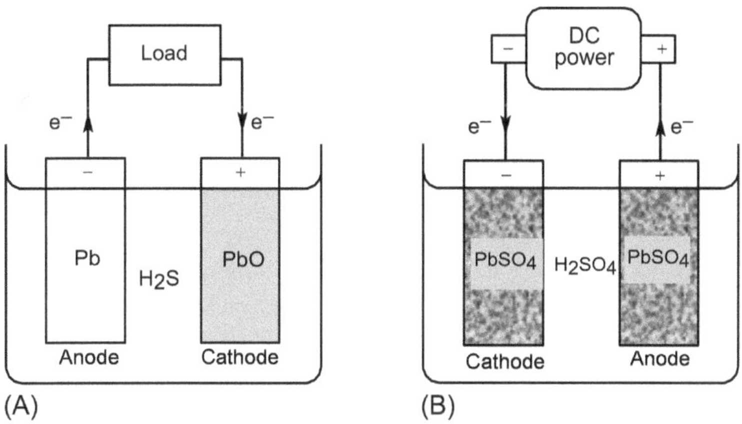

Fig. 6.3 Two stages of a lead-acid cell: (A) fully charged and supplying current to an external load and (B) fully discharged and receiving current from an external power source.

negatively charged lead anode is oxidized to lead (II) sulfate, an insoluble white solid, by the half reaction

$$Pb(s) + H_2SO_4(aq) \leftrightarrow PbSO_4(s) + 2H^+(aq) + 2e^- \quad (6.1)$$

The cathode consists of lead (IV) oxide deposited on a lead plate. During discharge, the red-brown oxide is reduced to lead (II) sulfate by the reaction

$$PbO_2(s) + 2H^+(aq) + H_2SO_4(aq) + 2e^- \leftrightarrow PbSO_4(s) + 2H_2O(l) \quad (6.2)$$

The cathode has a positive charge as it consumes electrons. The overall discharge reaction is

$$Pb(s) + PbO_2(s) + 2H_2SO_4(aq) \leftrightarrow 2PbSO_4(s) + 2H_2O(l) \quad (6.3)$$

Sulfuric acid is consumed, and water is produced in the reaction. Hence, both the concentration and the density of the electrolyte drop steadily during discharge. The state of charge of a lead-acid battery can be conveniently determined by measuring the specific gravity of its electrolyte. The battery is acting as an electrolytic cell, while it is recharged by an external power source as shown in Fig. 6.3B. The two electrodes now have lead (II) sulfate deposited on their grids. With the electric current forced to flow in the opposite direction, the former anode functions as a cathode and vice versa. This change in roles occurs with any rechargeable cell.

The cell voltage is obtained by calculating the free energies of the chemical reactions and using the Nernst equation. The ideal free energy change for the reaction is calculated as $\Delta G^o = \Sigma \Delta G_f^o$ (products) $- \Sigma \Delta G_f^o$ (reactants), where ΔG_f^o is the ideal free energy of formation of each species. Using thermodynamic data, we find that $\Delta G^o = -371.1$ kJ at 298.15 K. The ideal voltage, or electromotive force, E_{cell}^o, of the cell can then be obtained from the relationship $\Delta G^o = -nFE_{cell}^o$, where $n = 2$ mol for this reaction and Faraday's constant, F, equals 96,485 C/mol. The result is $E_{cell}^o = 1.92$ V. In reality, voltage will be less than this value 'and is given by Nernst equation $E_{cell} = E_{cell}^o - (RT/nF)\ln Q$, where the reaction quotient Q contains the activities of all species in the reaction and depends on the acid concentration. For electrolyte with 0.1 molal H_2SO_4 $Q = 1 \times 10^4$. Then, we obtain $E_{cell} = 1.80$ V by substitution into the Nernst equation. The unit ideal voltages for main batteries are shown in Table 6.2 along with anode, cathode, and overall reactions. The cell typically delivers less voltage than the *ideal* cell because its acid concentration is lower.

The first generation of lead-acid batteries was mobile electrolyte type where electrolyte was circulated through the battery compartments containing immersed electrodes. The hydrogen and oxygen gases generated during overcharge had to be vented into the atmosphere. This process reduced water content in the acid, and therefore, regular water maintenance was required. These batteries often failed due to a number of problems including stratification of the acid, expansion of the positive plate, corrosion of electrodes, and partial or incomplete charging. The lead-acid battery was significantly improved by the end of twentieth century with the introduction of a valve-regulated design. The valve-regulated batteries would not spill the acid when inverted in contrast to flooded lead-acid batteries. In addition, the new design of the battery does not need maintenance, has a smaller footprint, has no hazard of acid spill, and has improved performance. These batteries have moderate specific energy (35 Wh/kg) and energy density (70 Wh/L) and have energy and power efficiencies at 75% and 90%, respectively.

The suitability of a rechargeable battery for any particular application is judged by reference to a set of its characteristic attributes, including cost, energy storage capacity, power, weight, volume, and useful life. The relative importance of these attributes varies greatly from one application to another. Lead-acid battery systems are used in both mobile and stationary applications. Their typical applications are emergency power supply systems, stand-alone systems, battery systems for mitigation of output fluctuations from wind power, and starter batteries in vehicles. Lead-acid batteries are standard batteries for automobile starting, lighting, and ignition tasks for

Table 6.2 Chemical Reactions and Single Unit Voltages of Main Batteries During Discharge

Battery Type	Chemical Reactions, Anode, Cathode and Overall	Unit Voltage
Lead-acid	$Pb + H_2SO_4 \leftrightarrow PbSO_4 + 2H^+ + 2e^-$	2.0 V
	$PbO_2 + 2H^+ + H_2SO_4 + 2e^- \leftrightarrow PbSO_4 + 2H_2O$	
	$Pb + PbO_2 + 2H_2SO_4 \leftrightarrow 2PbSO_4 + 2H_2O$	
Lithium-ion	$LiC_6 \leftrightarrow C_6 + Li^+ + e^-$	3.7 V
	$CoO_2 + Li^+ + e^- \leftrightarrow LiCoO_2$	
	$LiC_6 + CoO_2 \leftrightarrow C_6 + LiCoO_2$	
Sodium-sulfur	$2Na \leftrightarrow 2Na^+ + 2e^-$	2.08 V
	$xS + 2e^- \leftrightarrow 4S^{-2}$	
	$2Na + xS \leftrightarrow Na_2S_x$ ($x = 3–5$)	
Nickel-cadmium	$Cd + 2OH^- \leftrightarrow Cd(OH)_2 + 2e^-$	1.0–1.3 V
	$2NiOOH + 2H_2O + 2e^- \leftrightarrow 2Ni(OH)_2 + 2OH^-$	
	$Cd + 2NiOOH + 2H_2O \leftrightarrow Cd(OH)_2 + 2Ni(OH)_2$	
Nickel-metal hydride	$MH + OH^- \leftrightarrow M + H_2O + e^-$	1.35V
	$NiOOH + H_2O + e^- \leftrightarrow Ni(OH)_2 + OH^-$	
	$NiOOH + MH \leftrightarrow Ni(OH)_2 + M$	
Sodium nickel chloride	$Ni + 2NaCl \leftrightarrow NiCl_2 + 2Na^+ + 2e^-$	2.58 V
	$2Na^+ + 2e^- \leftrightarrow 2Na$	
	$Ni + 2NaCl \leftrightarrow NiCl_2 + 2Na$	

which low cost has been the most attractive factor. Batteries for automobile applications in internal combustion engine vehicles operate from a state of charge of about 90% and are required to provide power for engine cranking of up to 10 kW. This translates to a specific power of up to 600 W/kg, normally sustained for <1 s. Automobile batteries can provide a total energy throughput of about 100 times the nominal capacity, and an end-of-life condition is reached after 5000–10,000 cranking events (4–6 years in temperate climates). The end-of-life failure is due to either corrosion of the grid or shedding of the active material from the positive plate. This limited lifetime is tolerated because of the low cost of the replacement batteries. There are several types of lead-acid batteries including the flooded battery requiring regular topping up with distilled water, the sealed maintenance-free battery having a gelled/absorbed electrolyte, and the valve-regulated battery. Lead-acid battery has a low cost ($300–600/kWh) and a high reliability and efficiency (70%–90%). It is a popular storage choice for power quality, uninterruptible power source (UPS), and some spinning reserve applications. Its application for energy management, however, has been very limited due to its short cycle life (500–1000 cycles) and low energy density (30–50 Wh/kg) because of inherent high density of lead. Lead-acid batteries also have a poor low-temperature performance and therefore require a thermal management system. Nevertheless, lead-acid batteries have been used in a few commercial and large-scale energy management applications up to 40 MWh.

6.2.2 Lead-Carbon Batteries

The lead anode of the lead-acid batteries has been partially or fully substituted by carbon in the progressive designs. This substitution of the negative electrode with carbon has resulted in better performance of the batteries and reduced life-cycle cost when compared with the old-fashioned lead-acid batteries for stationary storage applications. Carbon black in small amount ranging from 0.15% to 0.25% has been incorporated into to the negative electrode to reduce the buildup of lead sulfate on the lead electrode plate. The accumulation of concentration of lead sulfate per cycle reduced from 0.1% to 0.05% and 0.03% when black carbon was incorporated with 3 times and 10 times beyond the base level, respectively. In addition, improved cycle life of the lead-acid battery has been found by increasing the fraction of a standard carbon black from 0.2% to 2.0 wt%. The mechanisms governing the improvements by the addition of carbon to the lead electrodes are (i) separation of individual crystals of lead sulfate that provides larger contact of the electrolyte for the termination stage of the recharge reaction and (ii) as a medium for enhancing the area of the electronically conducting surface accessible for the precipitation of lead.

A more drastic change to the carbon addition is to completely replace the negative lead electrode with one made from high-surface-area carbon in carbon-carbon ultracapacitors. The hybrid device is essentially a PbC asymmetrical capacitor or a hybrid electrochemical device that still employs the standard lead oxide positive electrode as shown in Fig. 6.4A. During charge and discharge, the positive electrode undergoes the same chemical reactions that occur in a conventional lead-acid battery, that is, the PbO_2 reacts with acid and sulfate ions to form $PbSO_4$ and H_2O. The main difference in the lead-carbon battery is the replacement of the lead negative electrode with a high-surface-area carbon electrode that does not undergo any chemical reaction at all. Instead, the energy storage is realized by a double-layer (non-faradic) storage, along with a possibly H^+ pseudocapacitance (faradic) storage expressed as

$$nC_6{}^{x-}\left(H^+\right)_x \rightarrow (\text{discharge})\, nC_6{}^{(x-2)-}\left(H^+\right)_{x-2} + 2H^+ + 2e^- \tag{6.4}$$

During discharge, H^+ stored in the carbon negative electrode in the fully charged state moves to the positive electrode where they are neutralized to form H_2O. The nucleation and growth of $PbSO_4$ are eliminated in the negative electrodes, leading to a prolonged cycle life. Thus, the cell has reduced acid concentration swings from the charged to discharged state, which reduces grid corrosion on the positive electrode and leads to a long life. Overall, the lead-carbon hybrid device offers substantial improvement

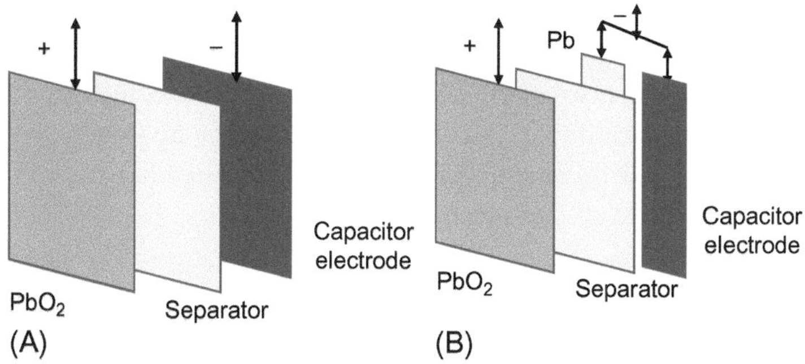

Fig. 6.4 Lead-carbon batteries (A) with lead electrode replaced by carbon capacitor electrode, (B) with half of lead electrode replaced by carbon capacitor electrode.

in energy output, compared with equivalent carbon-carbon ultracapacitors. In comparison with the traditional lead-acid battery, the lead-carbon asymmetrical capacitor is advantageous in power and cycle life because it completely avoids sulfate formation, although the discharge duration may be shortened. Therefore, the lead-carbon hybrid devices can be a promising storage technology for short period charge discharge applications.

An alternative to carbon addition or full replacement with a carbon electrode is a split design of the negative electrode where half of it is lead and the other half is carbon electrode called ultrabattery as shown in Fig. 6.4B. This design is composed of one lead-dioxide plate as the positive electrode and two negative plates, one a sponge lead negative plate (i.e., lead-acid anode) and another carbon-based negative anode (i.e., the asymmetrical capacitor anode). Essentially, the lead-acid negative plate and the ultracapacitor carbon electrode are internally connected in parallel to act as the anode of the hybrid battery. In such a device, the total discharge or charge current of the combined negative plate is composed of two components: the capacitor current and the lead-acid negative plate current. Accordingly, the capacitor electrode can now act as a buffer to share the discharge and charge currents with the lead-acid negative plate, and this prevents it from being discharged and charged at the high rates. The lead negative starts to convert to lead sulfate at a potential of about −0.98 V during discharge, and that converts back to sponge lead at a potential less than −1.0 V. On the other hand, the charge neutralization on the capacitor electrode occurs at a potential greater than −0.5 V during discharge, and subsequently, the charge separation occurs at a potential less than −0.3 V during charge. As such, during the early stages of discharge, the current would mainly come from the lead-acid negative plate and only little from the capacitor electrode owing to its higher charge-neutralization potential. During charging, the current flows to the capacitor electrode first and then to the lead-acid negative plate. This would lead to significant hydrogen evolution at the lead negative plate when it approaches toward the end of the charge. To solve the issue, the carbon electrode is modified via additives to such an extent that the hydrogen evolution on this electrode during charging has been reduced significantly to a level close to that of the lead-acid negative plate.

6.2.3 Electrochemical Performance and Challenges

Though lead-acid batteries have been widely used, there are main issues limiting development and expanded use of lead-acid batteries. There is progressive buildup of resistive lead sulfate layer on the electrode surface, with cycling, limiting the capacity and cycle life of lead-acid batteries. Lead-acid batteries have a low energy-to-weight ratio compared with other battery types, and they are slower to recharge than other battery types. With lead-carbon batteries, a total of 50% and 60% improvement in the discharge and charge power, respectively, have been observed. These hybrid batteries can be operated at a wider depth-of-discharge window but still can provide and receive similar power levels to that of the conventional lead-acid battery. The current ultrabattery is recommended for a relatively deep depth of discharge, 30%–70% (compared with that of <30% lead-acid battery), which has a slight self-discharge per week (1% with a relatively flat cell voltage of 2 V). There are demonstration efforts that are underway on the economic and technical viability of ultrabattery technology under American ARRA sponsorship. The challenge is coupling these batteries to the renewables energy storage where the slow ramp rate that would be a limit for uptake of wind or solar renewable energy and network stability. Furthermore, it remains unclear whether the ultrabattery offers a satisfactorily long deep-cycle life. The deep-cycle cells likely require thicker plates that may deliver less peak current, but they have to withstand frequent discharging. The lead-carbon asymmetrical capacitor and ultrabattery may be limited for applications that do not require long discharge duration. Increased duration would make the technologies available for not only power management but also energy applications. Scientifically, there is need of further basic study to understand the effects of carbon on electrode reactions, which involve the interaction and interfacial phenomenon between H^+ and high-surface-area carbon.

6.3 SODIUM-BETA ALUMINA MEMBRANE BATTERIES

Sodium (Na) is a favorable material to make anodes of batteries due to its abundant resources and low cost, along with its low redox potential. The Na anode is often separated with a cathode by a Na^+-conducting solid membrane in sealed electrochemical devices to avoid reaction with oxygen and water. A common membrane is beta-alumina that demonstrates an excellent Na^+ conductivity, particularly at elevated temperatures (300–350°C) to minimize electric resistance and achieve satisfactory electrochemical activities. The beta-alumina electrolyte shows a sufficient ionic conductivity ($\sim$0.2 s cm^{-1}) for battery applications only at temperatures of 250 °C or above. Sodium/beta-alumina batteries charge and discharge electricity via sodium ion transport across a beta-alumina electrolyte that is doped with Li^+ or Mg^{2+}. The anode is metallic sodium in a molten state during battery operation. The cathode can be either molten S/Na_2S_x, which is known as a sodium-sulfur (Na-S) battery, or solid metal halides or Zeolite Battery Research Africa (ZEBRA) batteries. The Na-beta batteries are commonly built in tubular designs. Structural components in tubular Na/S cells with central-sodium configurations include (1) cylindrical thin-walled beta-alumina tubes, closed at one end, encapsulating liquid sodium; (2) dense α-Al_2O_3 ceramic headers glass-sealed to open ends of beta-alumina tubes, providing electric insulation between anode and cathode compartments; and (3) tubular cathode containers fabricated from corrosion-protected metals or alloys, closed at one end, serving as current collectors for sulfur electrodes.

The poor sodium wetting is one of the most challenging issues in these batteries; it is related to the formation of a surface oxide film. Moisture absorbed on the electrolyte surface could react with sodium to form a surface layer of sodium oxide, which impedes sodium dissolution and sodium ion transport across the sodium anode/electrolyte interface. Impurities such as calcium in the electrolyte may also deteriorate the sodium wettability during long-term cell cycling. Calcium ions migrate to the interface and form a calcium-oxide film, which blocks contact between the sodium and electrolyte. The high operation temperature significantly increases battery manufacturing and maintenance costs and causes safety concerns as well.

6.3.1 Sodium-Sulfur and Sodium-Metal Halide Batteries

Sodium-sulfur (Na-S) batteries transform electric energy to chemical potential via electrochemical reactions shown in Table 6.2. During discharge phase, each sodium atom from the sodium metal electrode is stripped of an electron; the sodium ions thus formed migrate toward the positive electrode compartment through the electrolyte. The electrons freed from the sodium metal pass through the load circuit and return to the positive electrode. The returning electrons are captured by the molten sulfur to transform into polysulfide. The negative electron charge flow is balanced by the positively charged sodium ions migrating into the positive electrode compartment. This process is reversed during battery charging. Independent heaters that are part of battery system are used to keep the battery heated, and the battery temperature is maintained typically >300°C to facilitate the charge and discharge process. The sodium-sulfur battery yields a voltage of 1.78–2.208 V at 350°C, depending on the cell chemical reaction shown in Table 6.2 with $x = 3$–5. Sodium-sulfur batteries are highly efficient with efficiency typically at 89%.

Since the original development of the sodium-sulfur batteries by Ford Motor Company in the 1960s and subsequent acquisition by the Japanese company NGK and its collaborator Tokyo Electric Power Company, they have been employed for utility energy storage. A number of technology demonstrations have happened since 1990 in Japan. As a result of this, a great number of sodium-sulfur batteries have been manufactured and applied successfully in load leveling of energy storage systems and in electric vehicles. More than 270 MW of stored energy suitable for 6 h of daily peak shaving have been installed. The largest Na-S installation is a 34 MW, 245 MWh unit for wind stabilization in Northern Japan. The demand for Na-S batteries as an effective means of stabilizing renewable energy output and providing ancillary services is expanding. US utilities have deployed 9 MW for peak shaving, backup power, forming wind capacity, and other applications.

Sodium-metal halide batteries are built with a semisolid cathode that is made from porous metal/metal halide structures impregnated with molten $NaAlCl_4$ as a second electrolyte. Similar to the Na-S battery, the Na-metal halide batteries are commonly constructed on a tubular beta-alumina membrane. The energy conversion is carried out via the following reactions:

$$\left.\begin{aligned} \text{Anode}: 2\text{Na} &\leftrightarrow 2\text{Na}^+ + 2\text{e}^- \\ \text{Cathode}: \text{NiCl}_2 + 2\text{Na}^+ + 2\text{e}^- &\leftrightarrow \text{Ni} + 2\text{NaCl} \\ \text{Overall}: \text{NiCl}_2 + 2\text{Na} &\leftrightarrow \text{Ni} + 2\text{NaCl} \end{aligned}\right\} \tag{6.5}$$

The Na-halide batteries offer a standard voltage of 2.58 V at 300°C, slightly higher than that of Na-S batteries. During discharge, sodium ions are transported through the beta-alumina electrode from the anode to the cathode, reducing $NiCl_2$ to Ni via the migration of sodium ions in the molten $NaAlCl_4$ (eutectic of NaCl and $AlCl_3$) as the second electrolyte.

6.3.2 Components of Sodium Beta Batteries

Fig. 6.5A shows the schematic of the sodium-sulfur beta battery cell, and Fig. 6.5B shows the sodium-nickel chloride battery cell. The membrane of sodium-beta batteries is made from β_{00}-Al_2O_3 (beta-alumina), which is characterized with alternating, closely packed slabs and loosely packed layers. The loosely packed layers contain mobile cations (typically sodium) and are called conduction planes or slabs in which the cations are free to

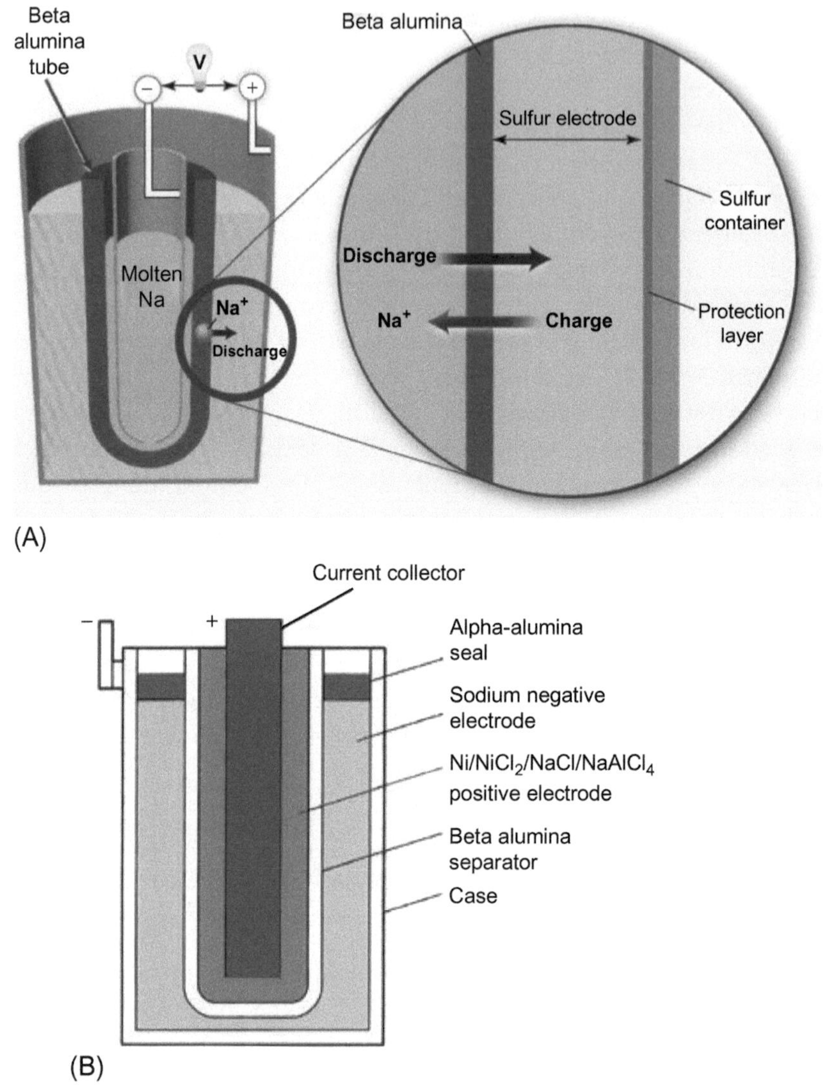

Fig. 6.5 (A) Tubular sodium-sulfur battery cell (B) sodium-nickel chloride battery cell.

move under an electric field. The closely packed slabs are layers of oxygen ions with aluminum ions sitting in both octahedral and tetrahedral interstices. These layers are referred to as a spinel block, which is bonded to two adjacent spinel blocks via conduction planes or slabs. The mobile cations diffuse exclusively within the conduction layers perpendicular to the conduction slab. There are two distinct crystal structures in the group:

β-Al_2O_3 hexagonal and β_{00}-Al_2O_3 rhombohedral. They differ in chemical stoichiometry and the stacking sequence of oxygen ions across the conduction layer. At 300°C, β_{00}-Al_2O_3 exhibits a Na^+ conductivity of, typically, 0.2–0.4 S/cm and is thus a favorable solid-state Na^+-conducting membrane. The substitution or doping allows higher sodium content and conductivity in β_{00}-Al_2O_3. Two favorable doping elements are Li and Mg. In real batteries, the ceramic electrolyte made of β_{00}-Al_2O_3 is added with zirconium dioxide. Water moisture tends to disintegrate the ceramic structure of β_{00}-Al_2O through infiltration and reaction at grain boundaries. The addition of zirconium dioxide to β_{00}-Al_2O_3 diminishes the sensitivity of pure β_{00}-Al_2O_3 to water moisture.

During cell operation, sodium in a liquid state serves as an anode for both sodium-sulfur and sodium-metal halide batteries. During discharge, the sodium is consumed resulting in the decrease of sodium volume, so it is important that the full active area of the beta-alumina solid electrolyte (BASE) is in contact with sodium. A good contact between the liquid sodium anode and the BASE is achieved by feeding sodium by gravity from a top reservoir, wicking sodium to the BASE surface, or forcing sodium from a reservoir by gas pressure. In tubular design of the battery, the wicking approach is used as it simplifies the sealing, enables high sodium utilization, and has overall compact design of the cell. The interfacial resistance between the BASE and the molten sodium has to be low and remain stable during entire life of battery operation. Complete wetting of the BASE by the melted sodium ensures low interfacial resistance. The impurities such as calcium can deposit as a film on interface of sodium and BASE, which hampers sodium ion transport and sodium dissolution. Thin layer coating of the BASE interface with lead and addition of titanium or aluminum into the liquid sodium can lessen the quantity of calcium oxidized at the interface.

During discharge of the sodium-sulfur battery, sodium polysulfides are formed when sulfur reacts with sodium ions, and during recharge, it is reformed. As an electronic conductor, carbon felt is typically inserted in the molten cathode to enable electron transfer since both sodium polysulfides and sulfur are electric insulators. The melts containing sulfur and polysulfides is very corrosive, resulting in the deposition of highly resistive impurities on the interface layer. Therefore, oxidization resilient materials, such as stainless steels along with a thin oxidization resilient layer of nonmetals such as carbon doped TiO_2 and $CaTiO_3$ are used. The cell failure mode is an issue or another concern with the sodium-sulfur battery. If the separating BASE is damaged, the liquid sodium comes in direct contact with sulfur, and polysulfides melts, and the chemical reactions between them is exothermic and characteristically dynamic, potentially causing fire and even explosion.

Metal halides are alternative cathodes to molten sulfur in sodium-metal halide batteries. For the sodium-metal halide battery, a secondary molten electrolyte such as $NaAlCl_4$ is used to serve as a sodium ion passage medium between the reaction zone in the cathode and the BASE, because most of the metal halides and the BASE are in solid phase. Moreover, the metal halide cathodes are preferred materials for current collector and cell case as they are less corrosive than sodium polysulfides. Ferrous chlorides have been widely used for cathode though nickel chloride is preferred since ferric chloride tends to form during overcharge, with ferrous chloride cathode. Sodium-metal halide batteries have another benefit over the sodium-sulfur chemistry, that is, they are safer in low-resistance cell failure mode.

Sodium/beta-alumina batteries have received much interest for transportation and stationary applications due to their high round-trip efficiency, high theoretical energy density, and extended cycle life. Wider market penetration requires several improvements to the battery. Cell performance improvements, economic fabrication through use of novel materials, unit cell and component design, and improved safety are required. For wide market penetration, the present capital cost for the battery is high at $500–600/kWh. Reducing operating temperatures of the battery improves materials durability, with the use of economic materials for stack and cell and less expensive thermal management.

6.4 NICKEL-CADMIUM AND NICKEL-METAL HYDRIDE BATTERY

6.4.1 Nickel-Cadmium Batteries

Nickel-cadmium batteries (NiCd) have well established in the market similar to lead-acid systems in terms of their maturity (100 years) and popularity. Nickel-based batteries have a higher power density and a slightly greater energy density (50–75 Wh/kg), and the number of cycles is higher (>3500 cycles) compared with lead-acid batteries. The NiCd batteries have nickel species and cadmium species as active materials of the positive and negative electrodes, respectively, and aqueous alkali solution KOH acts as the electrolyte. NiCd batteries usually have a metal case with a sealing plate equipped with a self-sealing safety valve. During the discharge cycle,

$Ni(OH)_2$ is the active material of the positive electrode, and $Cd(OH)_2$ is the active material of the negative electrode. During the charge cycle, NiOOH is the active material of the positive electrode, and metallic Cd is the active material of the negative electrode. NiCd battery is commonly found in two different forms, depending on the application: in its sealed form in portable equipment and in its flooded form in general industrial applications. The chemical reactions in the cell during discharge are shown in Table 6.2. A complication with this cell chemistry arises from the formation of oxygen gas at the cathode upon recharge of the cell. After ~80% charge, oxygen is generated via this formula.

$$2OH^- \rightarrow H_2O + \tfrac{1}{2}O_2 + 2e^- \qquad (6.6)$$

In a sealed cell, with a minimum amount of electrolyte, this oxygen can diffuse back to the anode, where it reacts to generate water and prevents any buildup of gas pressure. In a vented cell, also known as a pocket cell, the cell is flooded with excess electrolyte and includes a gas barrier to prevent this occurrence. The electrodes use an open nickel fiber or nickel foam structure impregnated with high-density spherical $Ni(OH)_2$. The positive and negative electrode plates, isolated from each other by the separator, are rolled in a spiral shape inside the case. The basic components consist of the positive and negative electrodes insulated by separators. The sandwiched electrodes are wound together and inserted into a metallic can that is sealed after injection of electrolyte. The batteries are typically sealed designs with metallic cases and tops that are electrically insulated from each other. The case serves as the negative terminal for the battery, while the top is the positive terminal. Finished battery designs may use a plastic insulating wrapper shrunk over the case to provide electric isolation between cells in typical battery applications. Fig. 6.6 shows the structure of the NiCd sealed batter cell.

In order to obtain a full-charge flooded NiCd battery, it is necessary to apply a certain level of overcharging, with a very quick charge cycle. The discharge cycle is also very quick, due to significant lower internal resistance. NiCd batteries can inject their rated power for 2 h. Large battery systems using vented NiCd batteries operate on a scale similar to lead-acid batteries. NiCd batteries are the only batteries capable of performing well even at low temperatures in the range from −20°C to −40°C.

Due to their robustness and low maintenance, they are favored for power tools, portable devices, emergency lighting, UPS, telecoms, and generator starting. For portable power applications, NiCd was the original mobile

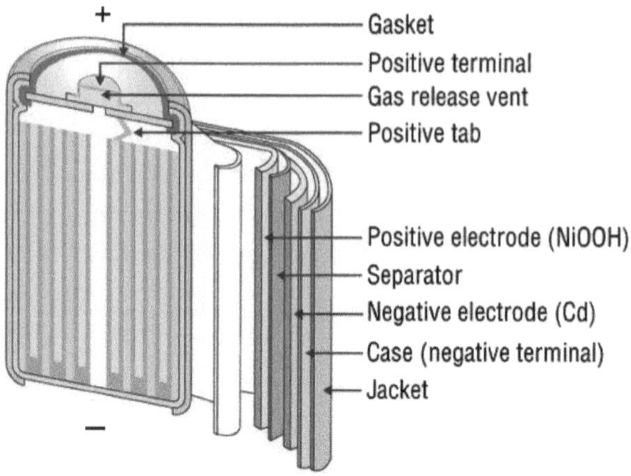

Fig. 6.6 Schematic of sealed spiral design of nickel-cadmium battery structure.

phone and notebook battery but lost out to Li-ion with its lighter weight and higher energy storage capability. Currently, it is the battery of choice for power tools and low-end electronics. It is characterized by an excellent high-rate and low-temperature capability and resistance to abuse and overcharge. NiCd batteries are used universally for starting and emergency power on jet aircraft because of their reliability and wide temperature operating range. Although the NiCd battery presents some good technical characteristics, it has not had a major commercial success, mainly due to its considerable costs (~\$1000/kWh) at >10 times of lead-acid batteries. Other major drawbacks of NiCd batteries are their toxicity and the fact that they suffer from the memory effect. Both cadmium and nickel are toxic heavy metals that can cause health risks.

6.4.2 Nickel-Metal Hydride Battery

Nickel-metal hydride batteries are similar to the proven sealed nickel-cadmium battery technology except that a hydrogen-absorbing negative electrode is used instead of the cadmium-based electrode. This eliminates cadmium, a toxic material, while this substitution increases the battery's electric capacity (measured in ampere-hours) for a given weight and volume. The chemical reactions during discharge are shown in Table 6.2. The negative electrode is made of hydride of metal alloys such as nickel and lanthanum alloy. The metal hydride electrode has a theoretical capacity >40% than the cadmium electrode in a nickel-cadmium couple. As a result, nickel-metal hydride batteries provide energy densities that are >20% than the equivalent nickel-cadmium battery. The electrolyte is alkaline, a 20%–40% weight solution of alkaline hydroxide containing other minor constituents to enhance battery performance. The baseline material for the separator is typically a nonwoven polyolefin, which provides electric isolation between the electrodes while still allowing efficient ionic diffusion. The nickel-metal hydride couple lends itself to the wound construction, which is similar to that used by cylindrical nickel-cadmium battery. Nickel-metal hydride batteries contain a resealable safety vent built into the top. The nickel-metal hydride battery is designed so the oxygen recombination cycle described earlier is capable of recombining gases formed during overcharge under normal operating conditions, thus maintaining pressure equilibrium within the battery. However, in cases of extended overcharge or incompatible battery charger combinations for the operating environment, it is possible that oxygen and hydrogen will be generated faster than it can be recombined. In such cases, the safety vent will open to reduce the pressure and prevent battery rupture. The vent reseals once the pressure is relieved. The expulsion of gas through the resealable vent can carry electrolyte, which may form crystals or rust once outside the can.

NiMH battery has moderate specific energy (70–100 Wh/kg) and relatively high energy density (170–420 Wh/L), significantly better than those of the NiCd battery. Other advantages of NiMH batteries over NiCd batteries include a reduced "memory effect," and they are more environment-friendly. NiMH batteries have the longer cycle life in comparison with Li-ion batteries. The NiMH battery has a wealth of applications from portable products to electric vehicles and potential industrial standby applications, such as uninterruptible power sources (UPS). The flat discharge characteristic, excellent high rate, long cycle life, and abuse tolerance have made NiMH the first choice for use in hybrid electric vehicles (HEVs). However, the significant barrier for HEV applications is the high rate of self-discharge, losing 5%–20% of its capacity within the first 24 h after fully charging. NiMH batteries currently cost about the same as lithium-ion batteries.

6.5 LITHIUM-ION BATTERIES AND APPLICATIONS

6.5.1 Lithium-Ion Batteries

Lithium-ion batteries, first proposed in the 1960s, came into reality when Bell Labs developed a workable graphite anode to provide an alternative to lithium metal (lithium battery). The first commercial lithium-ion batteries were produced by Sony in 1990. Since then, improved material developments have led to vast improvements in terms of the energy density (increased from 75 to 200 Wh/kg) and cycle life (increased to as high as 10,000 cycles). The efficiency of Li-ion batteries is almost 100%, which is an important advantage over other batteries. They have fast charge and discharge time constant defined as the time to reach 90% of the rated power of the battery, which is around 200 ms, with a relatively high round-trip efficiency of 78% within 3500 cycles. Lithium-ion batteries have become the most important storage technology in the areas of portable and mobile applications (e.g., laptop, cell phone, electric bicycle, and electric car) since around 2000. The discharge reactions and unit cell voltage are shown in Table 6.2. High cell voltage levels of up to 3.7 nominal volts mean that the number of cells in series with the associated connections and electronics can be reduced to obtain the required voltage. A lithium-ion cell can replace three NiCd or NiMH cells that have a cell voltage of only 1.2 V. The unit cells are made of anolytic and catholytic plates, filled with liquid electrolyte material, and the electrodes are separated by a porous separator of polyethylene or polypropylene, which allows the transit of lithium ions. The catholytic material is usually based on lithium metal oxide, such as $LiCoO_2$, $LiMO_2$, and $LiNiO_2$, while the anolytic material is graphite (C). The electrolytes used in lithium-ion cells are nonaqueous solutions, in which roughly 1 mol/dm^3 of lithium salts such as $LiPF_6$ or $LiClO_4$ is dissolved in a mixture of carbonate solvents selected from cyclic carbonates, such as ethylene carbonate and propylene carbonate, and linear carbonates, such as dimethyl carbonate, ethyl methyl carbonate, and diethyl carbonate. Charging and discharging cycles of the battery are detailed as follows. When the battery is charged, the lithium atoms in the cathode become ions and migrate through the electrolyte toward the carbon anode where they combine with external electrons and are deposited between the carbon layers as lithium atoms. This process is reversed during the discharge process.

Although Li-ion batteries have a share of over 50% of the small portable device market, there are some challenges for making large-scale Li-ion batteries. The main obstacle is the high cost (\$900–1300/kWh) due to special packaging and internal overcharge protection circuits. Since lithium-ion batteries are currently still expensive, they can only compete with lead-acid batteries in those applications that require short discharge times (e.g., as primary control backup). Safety is a serious issue in lithium-ion battery technology. Most of the metal oxide electrodes are thermally unstable and can decompose at elevated temperatures, releasing oxygen that can lead to a thermal runaway. To minimize this risk, lithium-ion batteries are equipped with a monitoring unit to avoid overcharging and overdischarging. Usually, a voltage balance circuit is also installed to monitor the voltage level of each individual cell and prevent voltage deviations among them.

Several companies are working to reduce the manufacturing cost of Li-ion batteries to capture large energy markets such the Lithium Battery Energy Storage Technology Association (LIBES), Kyushu Electric Power, and Mitsubishi Heavy Industries in Japan who have produced 3 kW demonstration modules. In the United States, the Department of Energy has sponsored a project by SAFT and SatCon Power Systems to design and construct two 100 kW/1 min lithium-ion battery energy storage systems for use in providing quality power for grid-connected microturbines. Both units are being tested at utility partner sites.

6.5.2 Applications of Li-Ion Batteries and Challenges

Key features such as high energy density and specific energy, 170–300 Wh/L and 75–125 Wh/kg, respectively, and their fast charge and discharge capability make Li-ion batteries good candidates for applications where the response time and weight are important. Lithium batteries have a long list of real-world applications from life-saving medical equipment to luxury yachts; lithium batteries keep both the essentials and the comforts of modern life running with safety and reliability. They have application in facilitating daily activities to providing essential emergency support. As emergency power backup or UPS, lithium batteries provide instant power to critical equipment power: computers, communication technology, and medical technology. Lightweight lithium batteries power recreational vehicle or electric vehicle with increased efficiency due to reduced weight and size as compared with lead-acid batteries. With a life span of over 10 years, lithium batteries provide power for long journeys and lose little power between uses.

Rechargeable lithium batteries are the best match for solar panels due to how they charge and how fast. Solar panels produce low-resistance charging, which is what lithium batteries require. In addition, lithium batteries charge quickly maximizing the potential solar power storage from each day of sunlight. Rechargeable lithium batteries are ideal for remote monitoring systems due to long life, small size, and not losing power via self-discharge during the time that your system is inactive. Lithium batteries have a self-discharge rate that is 10 times lower than lead-acid batteries, making them ideal for situations where they are not under continuous use. Lightweight lithium batteries are the ideal choice for mobility equipment, as they offer size customization, a longer life span, fast charging, a low self-discharge rate, and extended run time when compared with lead-acid batteries. Due to these qualities, the lithium-ion battery has recently been developed for stationary energy storage. Testing programs have been carried out for residential and community applications. Depending on the size of the energy storage, a number of battery pack units are combined for power levels ranging from few kilowatts to megawatts.

Given their high energy/power density and nearly 100% coulomb efficiency and anticipated mass production, Li-ion technologies have gained increasing interest for stationary applications. Several companies have experience in using Li-ion batteries in the utility-scale energy market. The US-based AES Energy Storage has been commercially operating a Li-ion battery energy storage (BES) system supplying frequency regulation for renewables including wind generation plants. Similar efforts are around the globe to integrate renewables into the grid. There are few challenges to lithium-ion batteries that are targeted for vehicle applications to deploy for stationary applications. The cost of the batteries developed for vehicle applications is two to five times higher on energy (kWh) basis than that for the stationary applications. The high material cost is the major contributor for this high cost per kilowatt-hour. There are also additional challenges regarding reliability and safety of lithium-ion batteries while using for stationary applications and for transportation applications. For electric grid applications where power levels are in megawatt scales, reliability and safety of the batteries become important. At this scale of power, the heat generated in the cells during charge and discharge cycle must be dissipated quickly to avoid overheating and enable effective use of the full capacity of the batteries.

The design of the battery for transport application have constrain on volume and weight preferably requiring high specific energy and high-energy-density storage. In contrast, for stationary applications, these qualities may not be strictly required. The low capital cost and longer battery life are important to the stationary application markets. It is prudent to develop cost-effective lithium battery technologies with long calendar and cycle life, before searching for an ideal lithium-ion battery that could meet the technical requirements for both transportation and stationary applications. For stationary applications, developing battery with cost-effective electrolytes and electrodes that can perform satisfactorily as an energy storage device at a low cost would be important even at the cost of sacrificing energy density and specific energy.

6.6 REDOX FLOW BATTERIES (RFB)

In Fig. 6.7, a schematic of redox flow battery (RFB) is shown for vanadium redox flow battery. RFB is a type of rechargeable battery that stores electric energy in external two electrolyte tanks typically containing soluble redox couples. The sizes of the tank are based on storage requirements for an application. Pumps are used to transfer the liquid electrolytes from storage tanks to battery compartments containing flow-through electrodes where oxidation and reduction processes convert chemical energy to electric energy during discharge and electrolytes gain chemical energy with supplied power during charging. The electrolytes flowing through the anode and cathode are referred to as anolyte and catholyte, respectively. There is a membrane (or separator) between the anode and cathode compartments that selectively allows cross transport of nonactive species (e.g., H^+ and Cl^-) to maintain electric neutrality and electrolyte balance. In contrast to conventional batteries, RFBs store energy in the electrolyte solutions. The power and energy ratings are independent of the storage capacity determined by the quantity of electrolyte used and the power rating by the active area of the cell stack. RFBs can release energy continuously at a high rate of discharge for up to 10 h. In comparison with fuel cells in which only the electroactive chemicals (e.g., hydrogen, methanol, and oxygen) flow through the reactor and the electrolyte remains at all times within the reactor, the electrolyte flows through the reactor in RFBs. RFBs are also distinguished from fuel cells by the fact that the chemical reaction involved is often reversible, i.e., they are generally of the secondary battery type, and so, they can be recharged without replacing the electroactive material.

The first redox flow batteries were the iron-chromium redox flow batteries developed at National Aeronautics and Space Administration. Since then, various types of RFBs have been developed that are named based on the chemistries of anolyte and catholyte. These RFBs include iron/chromium flow batteries (ICB), polysulfide-bromine flow batteries (PSBs), vanadium redox flow batteries (VRBs), zinc/bromine flow batteries (ZBB), vanadium-cerium flow batteries, and soluble lead-acid batteries. The RFBs, ICB, PSB, VRB, and ZBB have been built and operated with power ratings from few hundred kilowatts to multimegawatts.

6.6.1 All Vanadium Redox Flow Batteries

Among the FRBs, VRB is one of the most advanced flow battery systems with only redox couples (V^{2+}/V^{3+} and V^{4+}/V^{5+}) in anolyte and catholyte

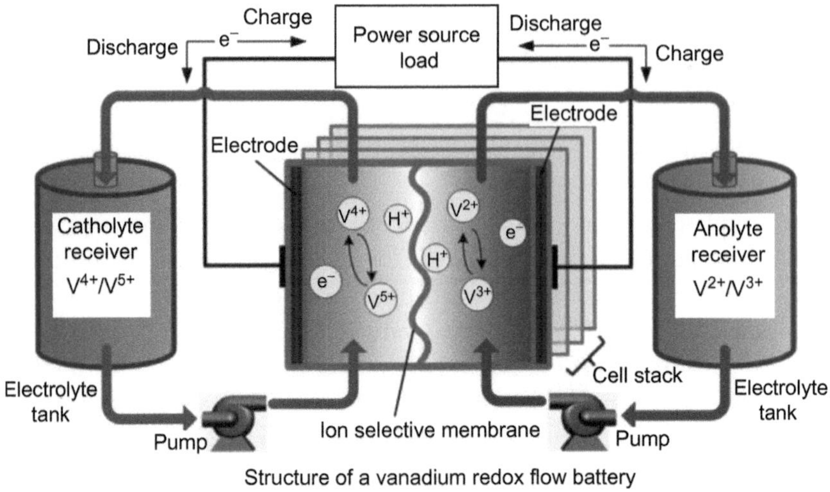

Fig. 6.7 All vanadium redox flow battery with catholyte and anolyte tanks.

electrolyte tanks. These are stored in mild sulfuric acid solutions (electrolytes). Due to the solubility of $VOSO_4$, a starting electrolyte, and the stability of vanadium species, the concentration of total $SO_4{}^{2-}$ and vanadium is usually controlled at <5 and 2 M, respectively. The solution of V^{4+} ions is made by dissolving the starting electrolyte $VOSO_4$ in sulfuric acid solutions. Having a single active element significantly diminishes cross contamination of the anolyte and catholyte. During the charge/discharge cycles, H^+ ions are exchanged between the two electrolyte tanks through the hydrogen-ion-permeable polymer membrane. During charging operation, the V^{3+} is converted to V^{2+} ions at the negative electrode, while at the same time at the positive electrode, V^{4+} ions are converted to V^{5+} ions through the release of electrons:

$$\left.\begin{aligned} \text{Cathode}: VO_2{}^+ + 2H^+ + e^- &\rightarrow (\text{discharge})VO^{2+} + H_2O \\ \text{Anode}: V^{2+} - e^- &\rightarrow (\text{discharge})V^{3+} \\ \text{Overall}: VO_2{}^+ + V^{2+} + 2H^+ &\rightarrow (\text{discharge})VO^{2+} + V^{3+} + H_2O \end{aligned}\right\} \quad (6.7)$$

These reactions absorb the electric energy and convert it to chemical energy. During the discharge phase, the reactions are reversed, which leads to the release of the stored chemical energy to electric energy. A single cell of VRB generates standard voltage of 1.26 V at 25 C. For higher voltage, the single cells are stacked in series using a pair of bipolar plates for each unit. The electrodes are often directly united with a bipolar plate as one

Table 6.3 Technical Features of Redox Flow Batteries

RFB Type	Open-Circuit Voltage (V)	Specific Energy (Wh/kg)	Characteristic Discharge Time (h)	Cycle Life (Cycles)	Round-Trip DC Energy Efficiency (%)
VRB	1.4	15	4–12	5000	70–80
ICB	1.18	<10	4–12	2000	70–80
PSB	1.5	20	4–12	2000	60–70
ZBB	1.8	65	2–5	2000	65–75

component to reduce electric contact resistance. In Table 6.3, key features of VRB, ZBB, ICB, and PSB are compared.

VRB can achieve an efficiency of about 85%. Their advantages include low maintenance cost, tolerance to overcharging, and ability to be deep charged without affecting the cycle life. On the other hand, the need for pumps, sensors, power management, and secondary containment makes them unsuitable for small-scale energy storage application. As an electric energy storage system, VRBs have been demonstrated up to multimegawatt power and megawatt-hour energy for intermittent renewable energy.

6.6.2 Other RFB

6.6.2.1 Iron/Chromium Flow Batteries (ICB)

The ICB is the earliest storage device invented in the 1970s at NASA that employs the redox couples of Fe^{3+}/Fe^{2+} and Cr^{3+}/Cr^{2+} in hydrochloric acid solution medium as catholyte and anolyte, respectively. The electrode and cell reactions are as follows:

$$\left.\begin{aligned} \text{Cathode}: Fe^{3+} + e^- &\rightarrow (\text{discharge})Fe^{2+} \\ \text{Anode}: Cr^{2+} - e^- &\rightarrow (\text{discharge})Cr^{3+} \\ \text{Overall}: Fe^{3+} + Cr^{2+} &\rightarrow (\text{discharge})Fe^{2+} + Cr^{3+} \end{aligned}\right\} \quad (6.8)$$

During the discharge cycle, Cr^{2+} is oxidized to Cr^{3+} in the negative half-cell, and an electron is released to do work in the external circuit through the negative and positive terminals of the AC/DC converter. In the positive half-cell during discharge, Fe^{3+} accepts an electron from the external circuit and is reduced to Fe^{2+}. These reactions are reversed during charge, when current is supplied from the external circuit through the AC/DC converter. Hydrogen (H^+) ions are exchanged between the two half-cells to maintain charge neutrality as electrons leave one side of the cell and return to the other side. The hydrogen ions diffuse through the separator, which electronically separates the half-cells. The cell reaction gives a standard voltage of 1.18 V. The ICB operates with either a cation or anion exchange membrane/separator and typically employs carbon fiber, carbon felt, or graphite as electrode materials. The Fe^{3+}/Fe^{2+} redox couple exhibits a very high reversibility and fast kinetics on the carbonaceous electrodes (carbon or graphite), whereas the Cr^{3+}/Cr^{2+} redox couple shows a relatively slow kinetics with the electrode materials. The hydrogen evolution appears to be a competitive reaction to the Cr^{3+}/Cr^{2+} anode during charging. Hence, catalysts Au, Pb, Tl, and Bi were employed for the Cr^{3+}/Cr^{2+} redox couple to enhance its electrode kinetics and to mitigate hydrogen evolution during the reducing process of Cr^{3+} to Cr^{2+}. Mixed electrolytes at both the cathode and anode sides are used to reduce cross transport of iron and chromium active spices. The mixed electrolytes allowed for the use of a cost-effective microporous separator, leading to a reduction in resistance.

6.6.2.2 Polysulfide/Bromine Flow Batteries (PSBs)

Sodium polysulfide and sodium bromide salt solutions are used as electrolytes in PSB system. The discharge reactions in PSB are as follows:

$$\left.\begin{aligned} \text{Cathode}: Br_3^- + 2e^- &\rightarrow (\text{discharge})3Br^- \\ \text{Anode}: 2S_2^{2-} - 2e^- &\rightarrow (\text{discharge})S_4^{2-} \\ \text{Overall}: Br_3^- + 2S_2^{2-} &\rightarrow (\text{discharge})3Br^- + S_4^{2-} \end{aligned}\right\} \quad (6.9)$$

The significant advantages of PSB systems are the materials of two electrolytes that are abundant, cost-effective, and highly soluble in aqueous electrolytes. The voltage generated across the membrane is 1.5 V; the PSB system has a fast response time, reacting within 20 ms. At the cathode side, the bromide ions are oxidized to bromine and converted as tribromide ions, and at anode side, the polysulfide anion is reduced to sulfide ion during the charging cycle. The reducing and oxidizing agents are sulfide ion and the tribromide ion, respectively. The electrolyte solutions are separated by a cation-selective membrane such as Nafion membranes to prevent the sulfur anions from reacting directly with the bromine, and the electric balance is achieved by transporting Na^+ across the membrane. Electrodes are made from materials such as high-surface-area nickel foam, carbon/graphite, and sometimes sulfide nickel.

PSBs have a wide range of potential application areas, especially for power system frequency control and voltage control due to their fast response characteristic. Because bromine and sodium sulfate crystals are produced during the chemical reactions, this may result in environmental issues. Several PSB systems have been demonstrated at multikilowatt scales. For instance, a 100 kW stack using PSB technology had been built by the UK company Innogy, with a net efficiency of 75%.

6.6.2.3 *Hybrid Flow Battery (HFB)*

In contrast to redox flow batteries, in a hybrid flow battery (HFB), one of the active masses is internally stored within the electrochemical cell, whereas the other remains in the liquid electrolyte and is stored externally in a tank. Thus, hybrid flow cells combine features of conventional secondary batteries and redox flow batteries: the capacity of the battery depends on the size of the electrochemical cell. Typical examples of a HFB are the zinc cerium and zinc-bromine flow battery systems. In both cases, the anolyte consists of an acid solution of Zn^{2+} ions. During charging, Zn is deposited at the electrode and at discharging Zn^{2+} goes back into solution. As membrane, a microporous polyolefin material is used; most of the electrodes are carbon-plastic composites.

6.6.2.4 *Zinc/Bromine Flow Batteries (ZBB)*

In each cell of a ZBB, two different electrolytes flow past electrodes in two compartments separated by a microporous polyolefin membrane. The electrodes are generally made from high-surface-area, carbon-based materials. Zinc-bromide aqueous solution with agents added is used as the electrolyte that is pumped through negative and positive electrode surfaces. The positive and negative electrodes are separated by microporous plastic film or ionic membrane as a separator. The ionic membrane selectively transfers bromide and zinc but not the polybromide ions, aqueous bromine, or complex phase. During discharging, bromide is converted to bromide ions at the positive cathode, and the process is reversed during charging. During these processes, the evolution of bromine is minimized by using complexing agents since bromine is a serious health hazard. The charge and discharge of the ZBB cells proceed via the following electrode reactions:

$$\left.\begin{aligned} \text{Cathode}: \mathrm{Br_2(aq)} + 2e^- &\rightarrow (\text{discharge})2\mathrm{Br}^- \\ \text{Anode}: \mathrm{Zn} - 2e^- &\rightarrow (\text{discharge})\mathrm{Zn}^{2+} \\ \text{Overall}: \mathrm{Br_2(aq)} + \mathrm{Zn} &\rightarrow (\text{discharge})2\mathrm{Br}^- + \mathrm{Zn}^{2+} \end{aligned}\right\} \quad (6.10)$$

During discharge, Zn and Br combine into zinc bromide, generating 1.8 V across each cell. This will increase the Zn^{2+} and Br^- ion density in both electrolyte tanks. During charge, metallic zinc will be deposited (plated) as a thin film on one side of the carbon-plastic composite electrode. Meanwhile, bromine evolves as a dilute solution on the other side of the membrane, reacting with other agents (organic amines) to make thick bromine oil that sinks down to the bottom of the electrolytic tank. It is allowed to mix with the rest of the electrolyte during discharge. The net efficiency of this battery is about 75%. Various companies are working on the commercialization of the ZBR hybrid flow battery, which was developed by Exxon in the early 1970s. In the United States, ZBB Energy and Premium Power sell trailer-transportable ZnBr systems with unit capacities of up to 1 MW/3 MWh for utility-scale applications. 5 kW/20 kWh systems for community energy storage are in development as well.

6.6.3 Challenges and Future R&D Needs for RFBs

Though RFB technology has advantage for energy storage and has successful demonstration of systems up to MWh levels, they have not enjoyed broad market penetration. Some of the reasons are relatively high capital and life-cycle costs. As an example, VRB has operating cost about $500/kWh or higher, which is obviously still too high for broad market penetration. The high cost of materials/components and performance parameters, including reliability, cycle/calendar life, energy efficiency, and system energy capacity, contribute to these high operating costs. The vanadium cost is high with the price fluctuating from $19 to $35 per kilogram (2017–18 annual low and high prices). Another expensive component is the Nafion-based membranes that also need further improvement in selectivity and chemical stability. The high reactivity of V^{5+} as a strong oxidant makes it challenging to select materials in terms of the long-term durability in VRBs. A number of improvements in RFB technologies such as design of unit cell, stack, and system and engineering besides materials/components and electrolyte chemistries are essential for improved performance and economy. Parasitic or shunt currents that lead to self-discharge and energy loss are some of the major issues in operating RFBs. Since all anode or cathode sides of the cells are supplied with pumped electrolyte in parallel, the current loss occurs. The shunt current is created due to voltage difference over different cells as the conductive electrolyte is fed commonly to the cells. In order to improve performance parameters, minimize the shunt current, and reduce overall system life-cycle costs, design optimization is required. Issues in performance, reliability, and lifetime still remain for all RFBs. High solubility of bromine in the aqueous zinc-bromide electrolyte and formation of zinc dendrites upon deposition on electrodes are some of the issues that have plagued ZBB development. The zinc dendrite growth can be reduced with a uniform current distribution in the cell. The evolution of hazardous gases such as bromine is health and environmental concerns in the RFB such as ZBB. The current RFBs are mainly operated in an aqueous electrolyte. To avoid gas evolution, their operation voltage, and thus their energy density, is limited.

Fundamental studies on chemistry and kinetics are necessary for many RFB systems. The electrochemical reactions in an RFB system are usually more complicated as they involve reactions at or near the electrode surface, mechanisms of charge transport and crossover in the ionic exchange membrane, and behavior of active species in a flowing electrolyte environment. The operational parameters, including electrolyte concentration, additives, current density, and temperature, are all the influential factors contributing to the complexity of RFB chemistry. Integration of electrochemical reactors with other devices and unit processes (e.g., optical-electrochemical sensors and combined adsorption/electrochemical treatment of soluble contaminants) is also crucial in an RFB system.

Modeling and simulation are certainly of great importance to RFB technology to understand the effects of variations of the entire system and ancillary equipment. Other topics of interests are to examine the proposed mechanism of electrochemical and chemical reactions and their kinetics against the experimental data, avoid working in hazardous environments as many RFB systems utilize toxic species, provide simulated long-term reliability and durability test, and reduce the R&D cost. In addition, environmental compatibility and energy/material sustainability of RFB technologies have to be addressed.

6.7 ELECTROCHEMICAL CAPACITORS

6.7.1 Basic Principles

Electrochemical capacitors (ECs)—sometimes referred to as supercapacitor or ultracapacitor—do not use the conventional solid dielectric, but rather, they use electrostatic double-layer capacitance and electrochemical pseudocapacitance to store electric energy. Fig. 6.8 shows the main classification of the ECs. The electrostatic double-layer capacitors use carbon electrodes or derivatives to achieve the separation of charge in a Helmholtz double layer at

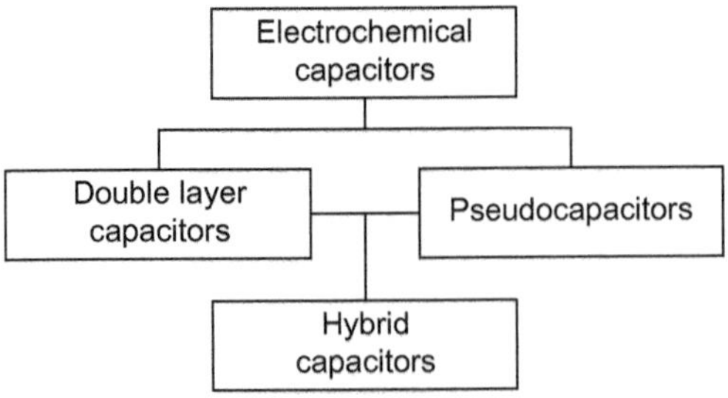

Fig. 6.8 Classification of electrochemical capacitors.

the interface between the surface of a conductive electrode and an electrolyte. The separation of charge is of the order of a 0.3–0.8 nm, much smaller than in a conventional capacitor. Electrochemical pseudocapacitors use metal oxide or conducting polymer electrodes with a high amount of electrochemical pseudocapacitance additional to the double-layer capacitance. Pseudocapacitance is achieved by Faradaic electron charge transfer with redox reactions, intercalation, or electrosorption. Hybrid capacitors also called asymmetrical capacitors, such as the lithium-ion capacitor, use two electrodes, one exhibiting mostly electrostatic capacitance with a capacitive carbonaceous electrode and the other mostly electrochemical capacitance with battery-like Faradaic electrode. This unique design can adjust the operating voltage window and increase the energy density due to the electrochemical redox process at the Faradaic electrode.

Electrochemical capacitors consist of two electrodes separated by an ion-permeable separator membrane and an electrolyte ionically connecting both electrodes. When the electrodes are polarized by an applied voltage, ions in the electrolyte form electric double layers of opposite polarity to the electrode's polarity. For example, positively polarized electrodes will have a layer of negative ions at the electrode-electrolyte interface along with a charge-balancing layer of positive ions adsorbing onto the negative layer. The opposite is true for the negatively polarized electrode.

In addition, depending on electrode material and surface shape, some ions may infiltrate the double layer becoming specifically adsorbed ions and contribute to the total capacitance of the EC. Electrochemical capacitors use the double-layer effect to store electric energy; however, this double layer has no conventional solid dielectric to separate the charges. Thus, two storage principles in the electric double layer of the electrodes that contribute to the total capacitance of an electrochemical capacitor are as follows: (i) double-layer capacitance, electrostatic storage of the electric energy achieved by the separation of charge in a Helmholtz double layer;

(ii) pseudocapacitance, electrochemical storage of the electric energy achieved by Faradaic redox reactions with charge transfer. Both capacitances are only separable by measurement techniques. The amount of charge stored per unit voltage in an electrochemical capacitor is primarily a function of the electrode size, although the amount of capacitance of each storage principle can vary extremely.

6.7.2 Electrostatic Double Layer Capacitors

As shown in Fig. 6.9, the electrostatic double-layer capacitance (EDLC) device has two double layers in series, one at each electrode-electrolyte interface. Charged double layer persists after power source is removed, providing a source of stored electric energy.

The actual specific double-layer capacitance of solid electrode-aqueous solution interfaces is on the order of a few tens of $\mu F/cm^2$, thus much higher than those found for conventional capacitors, and a function of the nature of both the electrode and the electrolyte solution. Porous carbon with very high specific areas, up to $2000\,m^2/g$, can be produced inexpensively. Assuming a capacitance of $10\,\mu F/cm^2$, a gram of such high-area carbon provides a capacity of 200 F, which could be matched only by a conventional capacitor of much larger dimensions. Advantage has been taken of this phenomenon to develop electrochemical capacitors based purely on carbon and suitable electrolytes with exceedingly high capacitances per unit weight or volume. Practically, these storage principles yield a capacitor with a capacitance value in the order of 1–100 F. As shown in Table 6.1, electrochemical capacitors have superb specific power compared with batteries but modest specific energies. This translates, in transportation terms, as good acceleration but poor range.

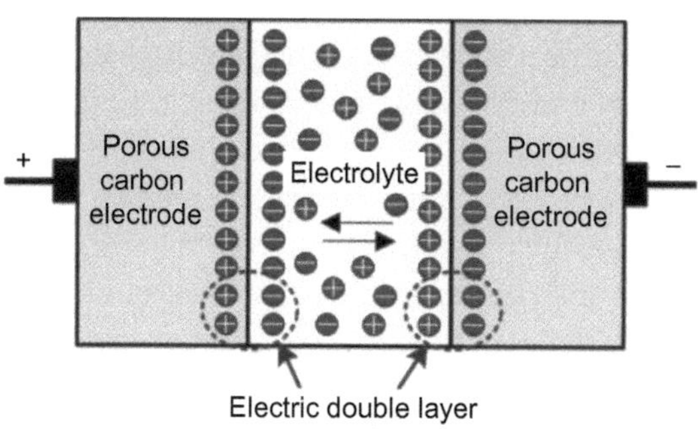

Fig. 6.9 Schematic representation of a charged electrochemical capacitor and double layers at both electrode-electrolyte interfaces.

The amount of charge stored per unit voltage in an electrochemical capacitor is primarily a function of the electrode size. The electrostatic storage of energy in the double layers is linear with respect to the stored charge and corresponds to the concentration of the adsorbed ions. Also, while charge in conventional capacitors is transferred via electrons, capacitance in double-layer capacitors is related to the limited moving speed of ions in the electrolyte and the resistive porous structure of the electrodes. Since no chemical changes take place within the electrode or electrolyte, charging and discharging electric double layers in principle are unlimited. Real electrochemical capacitor (EC) lifetimes are only limited by electrolyte evaporation effects.

6.7.3 Electrochemical Pseudocapacitors

In order to increase the capacitance of ECs, some electrochemical active redox material can be composited with the carbon material to make electrode materials for EC. In this way, the electron storage at the electrode/electrolyte interface of EDLC is not simply a physical process; some fast reversible oxidation/reduction reaction(s) will occur to give 10–100 times more capacitance than that of pure carbon-based EDLC. Applying a voltage at the electrochemical capacitor terminals moves electrolyte ions to the opposite polarized electrode and forms a double layer in which a single layer of solvent molecules acts as a separator. Pseudocapacitance occurs when specifically adsorbed ions out of the electrolyte diffuse the double layer. This pseudocapacitance stores electric energy by means of reversible faradaic redox reactions on the surface of electrodes in an electrochemical capacitor with an electric double layer. A transfer of electron charge occurs between electrolyte and electrode coming from a desolvated and adsorbed ion, whereby only one electron per charge unit is participating. This faradaic charge transfer originates by a very fast sequence of reversible redox, intercalation, or electrosorption processes. The adsorbed ion has no chemical reaction with the atoms of the electrode since only a charge transfer take place. The electrons involved in the faradaic processes are transferred to or from valence electron states of the redox electrode reagent. They enter the negative electrode and flow through the external circuit to the positive electrode where a second double layer with an equal number of anions has formed. The electrons reaching the positive electrode are not transferred to the anions forming the double layer; instead, they remain in the strongly ionized and transition metal ions of the electrode's surface. Materials exhibiting redox behavior for use as electrodes in pseudocapacitors are transition metal oxides like RuO_2, IrO_2, or MnO_2 inserted by doping in the conductive electrode material such as active carbon and conducting polymers such as polyaniline or derivatives of polythiophene covering the electrode material.

The storage capacity of faradaic pseudocapacitance is limited by the finite quantity of reagent in the available surface. A faradaic pseudocapacitance only occurs together with a static double-layer capacitance, and its magnitude may exceed the value of double-layer capacitance for the same surface area by factor 100, depending on the nature and the structure of the electrode because all the pseudocapacitance reactions take place only with desolvated ions, which are much smaller than solvated ion with their solvating shell. The ability of electrodes to accomplish pseudocapacitance effects by redox reactions, intercalation, or electrosorption strongly depends on the chemical affinity of electrode materials to the ions adsorbed on the electrode surface and on the structure and dimension of the electrode pores. The amount of electric charge stored in a pseudocapacitance is linearly proportional to the applied voltage.

6.7.4 Applications of Electrochemical Capacitors

Rechargeable lithium-ion batteries have dominated the energy storage market for decades. However, ECs have been recognized to provide improved energy storage capability for manifold applications from portable electronics to HEVs due to their high power density, rapid charge/discharge capability, and long life cycle. New technologies are being developed to optimize the performance and simultaneously reduce the cost of production of ECs for a number of applications. One of the first consumer products was the cordless screwdriver solely powered by ECs. ECs have been commercialized to power various house hold tools, remote controllers, and portable speakers. For industrial applications, ECs are generally employed as an emergency backup power source because of their instant discharging capability. In the event of a power failure in computer components, hospitals, and factories, ECs can assure a supply of uninterruptable energy to prevent catastrophic failures until the power is restored. Also, ECs are implemented as an emergency power source in airplanes. Unlike batteries, inert carbon-based ECs are stable in extreme conditions. This advantageous property created an opportunity to develop supercapacitor-powered drills for oil, petroleum, and geothermal exploration. EC-coupled hybrid devices serve as a backup power source for medical imaging equipment.

As a clean and alternative energy source, the wind energy market is constantly expanding, presenting more opportunities for ECs. ECs are capable of responding instantly to unpredictable weather conditions, supplying short bursts of power to operate the turbines. In addition, the low maintenance cost and long cycle life of ECs are suitable to be installed in unmanned wind turbines. Other applications of ECs have been found with airline cable cars and stand-alone, environment-friendly LED street lamps. ECs have been employed for storing regenerative breaking in public transportation vehicles.

6.8 CHEMICAL ENERGY STORAGE SYSTEMS

Chemical energy is stored in the chemical bonds of atoms and molecules, which is released when a chemical reaction occurs, and the substance is often changed into entirely different substance. Currently, chemical fuels are the dominant form of energy storage both for electric generation and for transportation. Coal, gasoline, diesel fuel, natural gas, liquefied petroleum gas (LPG), propane, butane, ethanol, biodiesel, and hydrogen are the most common chemical fuels that are processed. These chemicals are freely converted first to thermal and mechanical energy and then to electric energy by using heat cycles. The stored chemical energy can also be released through electron transfer reactions for the direct production of electricity using electrochemical devices such as fuel cells. Chemical energy storage is rather suitable for storage of large amounts of energy and for greater durations. Fig. 6.10 shows the specific energy, i.e., energy per mass or gravimetric density, and energy density or energy per volume or volumetric density for hydrogen and other chemical energy storage fuels based on lower heat values. For hybrid energy systems, the role of renewable being emphasized the chemical energy storage plays a major role. For such storage application, hydrogen and synthetic natural gas (SNG) serve as secondary energy carriers for primary energy derived from nonfissile primary energies nuclear and renewables including biomass. In addition to the conventional chemical fuels, new chemical and thermochemical energy storage technologies include sorption and thermochemical reactions such as ammonia system. The main purpose of large chemical energy storage system is to use excess electricity and heat to produce energy carrier, either as pure hydrogen or as SNG. Although the overall efficiency of hydrogen and SNG is low compared with storage technologies such as pumped hydro and Li-ion, chemical energy storage is the only concept that allows storage of large amounts of energy, up to the terawatt-hour range, and for greater periods of time—even as seasonal storage. These universal energy carriers can be used in different sectors, such as transport, mobility, heating, and the chemical industry.

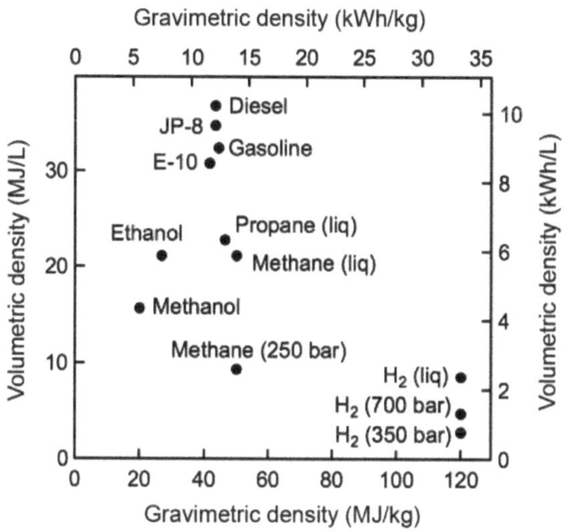

Fig. 6.10 Specific energy and energy density of hydrogen and several fuels based on lower heating values. *(Reproduced from https://www.energy.gov/eere/fuelcells/hydrogen-storage.)*

6.8.1 Hydrogen Energy Storage System

Hydrogen is a clean, highly abundant, and nontoxic renewable fuel and an ideal energy carrier material. The concept of a widespread hydrogen economy has been known more than century ago when it was suggested in Jules Verne's 1874 novel *The Mysterious Island*. Hydrogen is a widely used industrial chemical and can be produced from any primary energy source such as from water by thermolyses and electrolyses, reforming of fossil fuels, gasification of biomass, methanol, etc. It can be produced from water electrolysis using renewable electricity or produced directly from sunlight by photocatalytic water splitting. The hydrogen production methods have been discussed in detail in Chapter 4, Nuclear Hydrogen Production. Chemical energy of hydrogen is 142 MJ/kg (high heating value), which is higher than other hydrocarbon-based fuel, and its only combustion product is water. Hydrogen can be used to drive combustion turbines or fuel cells, in hydrogen cars with fuel cells or special internal combustion engines or for heat generation. Contemporary fuel cells can achieve 50%–55% efficiency, and a similar efficiency can be achieved with a gas and steam turbine.

Hydrogen storage methods can be divided in two main categories as physical (in gas or liquid phase) and material-based storage. Fig. 6.11 shows the storage methods for hydrogen. Gas-phase storage is generally made as

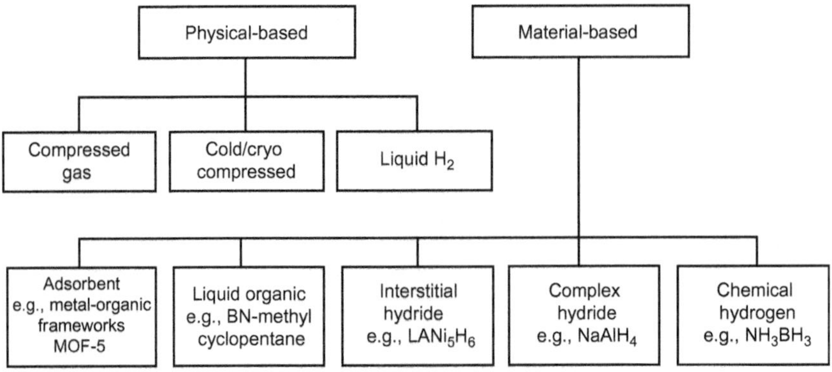

Fig. 6.11 Hydrogen storage methods.

compressed gas in high-pressure tank of 350–900 bar. Boiling point (at 1 atm.) of hydrogen is −252.8°C requiring liquid storage of hydrogen by cryogenic cooling methods. Material-based storage is through metal hydride, chemical hydrogen storage, and sorbent materials. It can also be stored on the surfaces of solids by adsorption or within solids by absorption. Material-based storage technology include metal-organic frameworks, metal hydrides, aluminum hydride, sodium alienate, magnesium hydride, and reactive hydride composite of $LiBH_4$ and MgH_2. Smaller amounts of hydrogen can be stored in aboveground tanks or bottles under pressures up to 900 bar. For larger amounts of hydrogen, underground piping systems or even salt caverns with several 100,000 m^3 volumes under pressures up to 200 bar can be used. Additional energy is required for compression duty corresponding 10%–15% of the energy content in the hydrogen. A higher energy density can be achieved by liquefying the hydrogen, but the cooling duty required for the liquefaction corresponds to about 30% of the energy contained in the hydrogen.

A typical hydrogen energy storage system consists of a hydrogen generation unit based on primary energy sources preferably nonfossil (renewable and nuclear), hydrogen storage system, hydrogen supply system, conversion of hydrogen to electricity, and electricity supply system to consumer. Fig. 6.12 shows the concept of hydrogen energy storage system. The generation of hydrogen can have different routes, low-temperature electrolysis, high-temperature electrolysis, or thermochemical or electrochemical processes to split water into hydrogen and oxygen using electricity and/or heat energy. Hydrogen is stored under pressure in gas bottles or tanks or as liquid hydrogen and in large salt caverns, and the storage can be done practically for

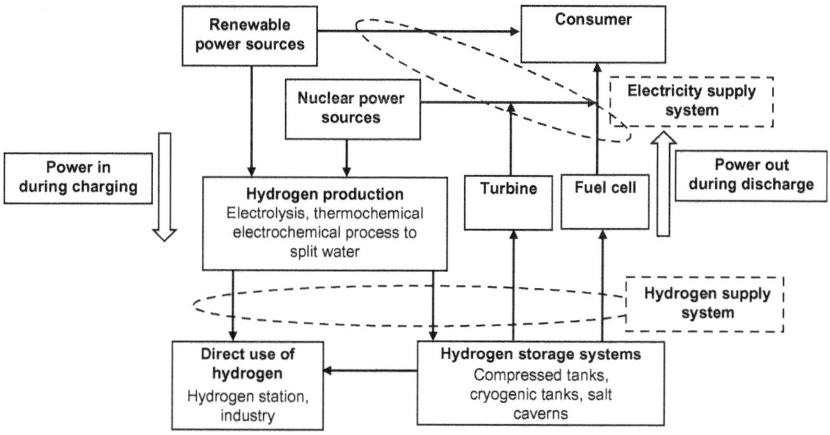

Fig. 6.12 Hydrogen energy storage technology.

an unlimited time. The hydrogen supply systems can consist of gas piping, cryogenic piping, transport trucks, and railways. To generate electricity, both gases flow into the fuel cell where an electrochemical reaction that is the reverse of water splitting takes place: Hydrogen and oxygen react and produce water, heat is released, and electricity is generated. For economic and practical reasons, oxygen is not stored but vented to the atmosphere on electrolysis, and oxygen from the air is taken for the power generation. In addition to fuel cells, gas motors, gas turbines, and combined cycles of gas and steam turbines can be used for power generation. The hydrogen systems with fuel cells (<1 MW) and gas motors (under 10 MW) can be adopted for combined heat and power generation in decentralized installations. Gas and steam turbines with up to several hundreds of megawatts could be used as peaking power plants. The overall AC-AC efficiency is around 40%. Similar concepts equally apply to other chemical energy storage fuels such as methane, SNG, and methanol.

6.8.2 Fuel Cell

A fuel cell is an electrochemical device that uses reverse electrochemical reactions of an electrolyzer and continuously converts chemical energy content of the fuel into electric energy, water, and some heat as long as fuel and oxidant are supplied. Fuel cells work like batteries, except that the electricity is produced as long as fuel is supplied, and do not run down or need recharging. When electricity is required, fuel and oxidant are fed to the cell continuously and simultaneously; the reaction products are removed.

Theoretically, electricity will be produced indefinitely if the supply and removal of materials is maintained. When hydrogen is used as a fuel, it generates only electric power and pure water and so referred to as a zero-emission engine. A fuel cell consists of two electrodes separated by an electrolyte medium as shown in Fig. 6.13. The electrolyte serves as a medium for ion conduction and avoids a direct contact of hydrogen and oxygen and a direct chemical combustion. Electrodes are usually composed of a gas diffusion layer with thin catalyst coatings at the electrode-electrolyte interfaces. Hydrogen and oxygen reactants are supplied to the anode and cathode electrode's surfaces, respectively. The electrodes are generally in the form of flat or circular annular surfaces to increase the contact surface area for reactions. The structure of the electrode is made porous for easy transport of reactant gases toward electrolyte. This also leads to higher contact area between the gas, the electrode, and the electrolyte.

In a fuel cell, a direct hydrogen-oxygen combustion reaction is replaced by two electrochemical half-reactions, oxidation and reduction at two electrode-electrolyte interfaces. The hydrogen and oxygen reactants are spatially separated by the electrolyte. The electrons released from breaking the hydrogen bond transfer through an external circuit to recombine with oxygen and form water as the new hydrogen-oxygen bond. The electrochemical reactions at the anode and cathode sides take place simultaneously producing electricity, water, and heat as the only by-products when hydrogen is used as the fuel. Heat is produced due to the irreversibilities associated

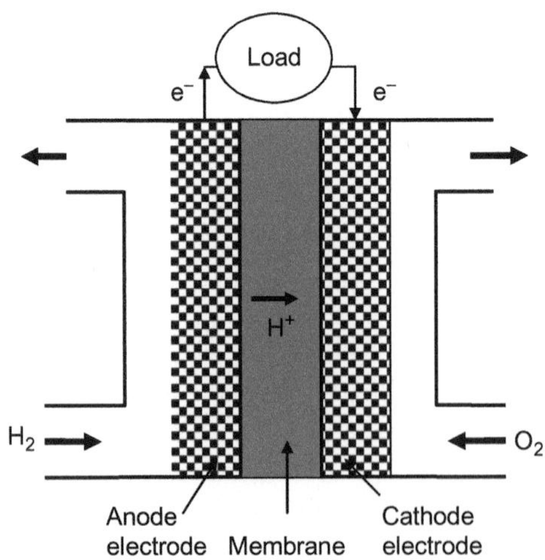

Fig. 6.13 Schematic fuel cell with hydrogen-oxygen as reactants.

with the electrochemical reactions at electrodes and by ohmic heating caused by the charge transport through cell components.

In one of the typical fuel cells, for example, a fuel (usually hydrogen in a hydrogen fuel cell) undergoes an electrochemical oxidation reaction and transforms into hydrogen ion or proton by releasing electrons at the anode. The charged ions or protons transport through the ion-conducting but electronically insulating electrolyte material from anode side to cathode side. At the cathode, oxygen undergoes electrochemical reduction reaction by combining with the incoming protons and electrons, producing water. The electrons flow through the electrically conducting electrodes and the external load circuit, resulting in electricity and performing electric work. The two electrochemical half-reactions and the overall reaction that represents the indirect combustion of hydrogen in the fuel cell are as follows:

$$\left.\begin{aligned} \text{Anode reaction}: H_2 &\rightarrow 2H^+ + 2e^- \\ \text{Cathode reaction}: \frac{1}{2}O_2 + 2H^+ + 2e^- &\rightarrow H_2O \\ \text{Overall reaction}: H_2 + \frac{1}{2}O_2 &\rightarrow H_2O \end{aligned}\right\} \quad (6.11)$$

A number of different types of fuel cells have been developed and are deployed for a variety of applications. They are generally classified or characterized primarily by the type of electrolyte used, the type of ion transferred, and the range of applicable temperature level. Table 6.4 shows a list of most commonly used fuel cells along with the type of electrolyte used, migrating ions, operating temperature ranges, and type of fuel used. Fuel cell power generation systems are considered for a wide range of applications including transportation, stationary power generation, portable power generation, and space and military applications.

6.8.3 Other Chemical Energy Storage Systems

6.8.3.1 Synthetic Natural Gas (SNG)

Natural gas is the second option to hydrogen to store electricity as chemical energy. Natural gas is most popular gas fuel, which mainly consists of methane (CH_4). Biogas, landfill gas, SNG, and bio-SNG are the other gas fuels. Biogas is produced by decayed organic matters and contains CH_4 and CO_2. Composition of landfill is similar to biogas. Synthetic natural gas (SNG) means the partly conversion of solid feedstock such as coal or wood, with gasification followed by gas conditioning, SNG synthesis, and gas upgrading

Table 6.4 List of Fuel Cells Classified Based on Electrolytes, Ions, and Temperature Range

Fuel Cell Type	Charge Carrier	Electrolyte Type	Typical Operating Temperature	Fuel and Oxidant
Alkaline fuel cell (AFC)	OH^+	Potassium hydroxide or sodium solution	$60-120^\circ C$	H_2, O_2
Phosphoric acid fuel cell (PAFC)	H^+	Phosphoric acid	$\approx 220^\circ C$	Pure H_2, O_2
Molten carbonate fuel cell (MCFC)	$CO_3{}^{2-}$	Lithium or potassium carbonate	$\approx 600-700^\circ C$	H_2, CO, CH_4 and other hydrocarbons, O_2
Proton exchange membrane fuel cell (PEMFC)	H^+	Solid polymer (Nafion)	$\approx 80^\circ C$	Pure H_2, O_2
Solid oxide fuel cell (SOFC)	O^{2-}	Solid oxide electrolyte (yttria-stabilized zirconia)	$\approx 700-1000^\circ C$	H_2, CO, CH_4 and other hydrocarbons, O_2
Direct methanol fuel cell (DMFC)	H^+	Solid polymer	$\approx 80^\circ C$	Methanol, O_2

or similar processes to natural gas. The SNG produced can be stored in pressure tanks, underground, or fed directly into the gas grid. The production of SNG is preferable at locations where both CO_2 and excess electricity are available. Steam-oxygen gasification, hydrogasification, and catalytic steam gasification are the different processes that could be used to convert coal to synthetic natural gas. Biomass could also be utilized for SNG production. The hydromethanation or catalytic steam gasification technology is considered more energy-efficient than the traditional methanation processes.

Synthesis of methane (also called synthetic natural gas, SNG) is the second option to store electricity as chemical energy. Here, a second step is required beyond the water splitting process in an electrolyzer, a step in which hydrogen and carbon dioxide react to methane in a methanation reactor. As is the case for hydrogen, the SNG produced can be stored in pressure tanks, underground, or fed directly into the gas grid. Several CO_2 sources are conceivable for the methanation process, such as fossil-fueled power stations, industrial installations, or biogas plants. To minimize losses in energy, transport of the gases CO_2 (from the CO_2 source) and H_2 (from the electrolysis plant) to the methanation plant should be avoided. The production of SNG is preferable at locations where CO_2 and excess electricity are both available. In particular, the use of CO_2 from biogas production processes is promising as it is a widely used technology. Nevertheless, intermediate on-site storage of the gases is required, as the methanation is a constantly running process. Methane has a higher energy density, and transport in pipelines requires less energy. The main disadvantage of SNG is the relatively low efficiency due to the conversion losses in electrolysis, methanation, storage, transport, and the subsequent power generation. The overall AC-AC efficiency, $<35\%$, is even lower than with hydrogen.

6.8.3.2 Methane

Methane (CH_4) is available as a natural gas and makes an interesting storage molecule. At identical pressures, the volumetric energy density of gaseous methane is more than three times the volumetric energy density of gaseous hydrogen. There is already a widespread infrastructure for transportation and handling of natural gas around the globe. Besides domestic use, gas stations for fueling vehicles running on natural gas are now appearing in various parts of the world. Methane can be produced from syngas by hydrogenation of CO and CO_2 typically with the use of a nickel catalyst. The methanation reaction is known for a long time and industrially used for the fine cleaning of the feed in ammonia plants:

$$CO + 3H_2 \rightarrow CH_4 + H_2O \tag{6.12}$$

$$CO_2 + 4H_2 \rightarrow CH_4 + 2H_2O \tag{6.13}$$

The recycling of CO_2 is an attractive proposition. The stored methane can be combusted to generate electricity in periods with insufficient renewable electricity production. The CO_2 produced in the methane combustion can then be hydrogenated into CH_4 with an estimated energy efficiency of around 75%–80%. The methane can then be stored in the natural gas grid and in salt caverns. This concept requires storage of significant amounts of H_2 and perhaps also CO_2 between methane synthesis and methane combustion. There are also challenges related to the use of gaseous methane as its energy density is lower than that of liquid fuel chemicals. There are safety aspects in the handling of methane, and methane has a greenhouse gas potential approximately 25 times (mol/mol) higher than CO_2, which means that leakages from a large-scale methane infrastructure must be avoided.

In comparison with hydrogen, both methane and hydrogen could be converted to electricity at approximately the same efficiency, either in gas and steam turbines or in fuel cells. The major advantages are infrastructure advantages. Methane could be fed into the natural gas pipeline system, and underground storage in caverns allows three times higher volumetric storage density for methane compared with hydrogen. However, if one would produce the hydrogen in a centralized electrolyzer near a gas cavern for buffering electric energy, it appears more effective to store the hydrogen directly instead of losing another 25% of the energy content for methanation. Energetically, it is much better to convert the hydrogen to electric energy and replace fossil fuel in electricity generation than capturing carbon dioxide from fossil fuel combustion, hydrogenating it, and then combusting the methane. However, methanation may be a viable option if the process would start from biomass. Overall efficiencies of gasification/methanation schemes are quoted to be in a similar range as the efficiencies of anaerobic fermentation, and thus, from an energy efficiency point of view, these technologies seem to be comparable. A disadvantage for the gasification/methanation schemes is most probably the consideration of economies of scale. While biogas plants can be relatively small, this is not the case for gasification/methanation plants, and the long transport distances for the biomass will then negatively affect the efficiency of such schemes.

6.8.3.3 Methanol, Ethanol, and Higher Alcohols

Methanol has the advantage of being a liquid at normal conditions of pressure and temperature, as our current energy infrastructure with its heavy reliance on oil largely is geared toward liquid fuels. This also opens the way for diverse industrial chemical applications ranging from olefin synthesis, such as polyethylene and polypropylene, to protein extraction in microalgae farming with other various applications such as improving the octane of traditional gasoline. Methanol can be produced from syngas by hydrogenation of CO and CO_2 at 300°C under 70 bar in a reactor with a metallic catalyst such as copper and zinc oxides on an alumina-based ceramic, $Cu/ZnO/Al_2O_3$. The ceramic is particularly adapted to this highly exothermic reaction. Along with the methanol synthesis reactions, the water-gas shift reaction takes place as well:

$$CO + 2H_2 \rightarrow CH_3OH \tag{6.14}$$

$$CO_2 + 3H_2 \rightarrow CH_3OH + H_2O \tag{6.15}$$

In the long term, methanol could also be used in vehicles powered by a direct methanol fuel cell, although research is needed, for example, to alleviate the need for expensive noble metal catalysts in the fuel cell. Methanol, which is easily transported, can also be reformed locally to produce hydrogen. Additionally, methanol can be converted into dimethyl ether. There are also well-established processes for converting methanol into synthetic gasoline. There are also challenges associated with the use of methanol. The volumetric energy density of liquid methanol is, although higher than that of hydrogen and gaseous methane, nevertheless considerably lower than that of higher alcohols and liquid hydrocarbons. In the presence of water, there are also issues with the miscibility of methanol and gasoline, which makes it difficult to phase in the fuel use of methanol. Compared with the other chemical storage molecules, there is no good and straightforward ethanol synthesis relying on purely chemical processes. But ethanol can be very easily obtained by fermentation routes and under favorable conditions. For example, yeasts convert fermentable sugars, such as glucose, to ethanol and carbon dioxide:

$$C_6H_{12}O_6 \rightarrow 2CH_3CH_2OH + 2CO_2 \tag{6.16}$$

This reaction occurs under anaerobic conditions, and depending on the type of yeast, concentrations of ethanol up to slightly above 20% can be achieved. Most current feedstocks for fermentation processes rely on crops that are also used for food and feed, such as sugar cane, sugar beet, wheat, or other cereals. As the use of oil crops for the production of hydrocarbons, this is not sustainable under most circumstances. Fermentation processes resulting in ethanol could become even more attractive, if lignocellulose instead of sugar- or starch-containing plants could be used as starting material for ethanol production. A mixture of methanol and higher alcohols (ethanol, propanol, and butanol) can also be prepared directly from syngas. In the presence of water, the mixed alcohols are more compatible with the existing infrastructure, since they have a better miscibility with oil-derived gasoline than pure methanol, and the mixed alcohol synthesis might therefore be advantageous during a transition to an energy infrastructure with methanol as a significant energy carrier. Ethanol could, just like methanol, be an important chemical for the chemical industry.

6.8.3.4 Liquid Hydrocarbons

Liquid hydrocarbons are the main components of current energy infrastructure, and liquid hydrocarbons produced from nonfossil energy sources, nuclear and specifically renewable resources, would thus be completely compatible with the existing energy infrastructure. The synthetic gasoline and diesel fuels have significantly higher energy densities than methanol, ethanol, methane, and hydrogen that will support aviation sector far into the future. Well-known Fischer-Tropsch process transforms coal into syngas ($CO+H_2$) and then into hydrocarbon fuels.

Liquid hydrocarbons can be produced from syngas in the well-known Fischer-Tropsch synthesis. The Fischer-Tropsch reaction, which occurs along with the water-gas shift reaction, can be expressed as

$$n(CO+2H_2) \rightarrow [CH_2]_n + nH_2O \tag{6.17}$$

Here, CH_2 represents the basic building block in the liquid hydrocarbons. Various gasoline-range hydrocarbons can also be produced from methanol and/or dimethyl ether over a zeolite catalyst with reactions

$$nCH_3OH \rightarrow [CH_2]_n + nH_2O \tag{6.18}$$

This synthesis of liquid hydrocarbons from syngas generally occurs with a lower thermal efficiency than the syntheses of, for example, methanol, dimethyl ether, or methane resulting in lower fraction of the primary energy stored in the synthesized hydrocarbon fuels. It is also possible to produce liquid hydrocarbon fuels directly from a catalytic conversion of the sugars in biomass or from treatment of pyrolysis-derived bio-oil with hydrogen. These processes require hydrogen. There are many biomass gasification schemes with subsequent Fischer-Trospch synthesis in pilot plant operation worldwide. In spite of the conceptually straightforward process, there are substantial challenges associated with it, which are primarily related to the syngas generation from a source with high levels of impurities and varying quality. The overall efficiency of hydrocarbon generation from renewable resources is only moderate compared with other pathways to possible fuels, but it is expected to play an important role in the future, because hydrocarbons have the highest gravimetric and volumetric energy density of possible fuels. They will probably be essential as fuels in air traffic and possibly also for heavy-duty trucks.

6.8.3.5 Ammonia

Ammonia, the second most produced chemical in the world (176 million tons in 2014), is manufactured at large plants (1000–1500 t/day) using Haber-Bosch process developed more than a 100 years ago. Modern NH_3-producing plants depend on hydrogen (H_2) using steam methane reforming (SMR) to react with atmospheric nitrogen (N_2) using a catalyst under high pressure and temperature (200 bar and 450°C) to produce anhydrous liquid NH_3. This step is known as the Haber-Bosch synthesis:

$$N_2 + 3H_2 \rightarrow 2NH_3 \tag{6.19}$$

With the global transition from fossil fuels to intermittent renewable energy sources, there is a need for long-term storage and long-range transmission of energy, for which ammonia is perfectly fit. To make it practical, it is necessary to match the scale of ammonia production with the scale of renewable energy sources, at the same or better capital cost per ton of NH_3, and reduce the energy consumption. The energy efficiency of the process (from natural gas) can be up to 70%. Due to the widespread use of ammonia as a fertilizer, there is already some infrastructure for transportation of ammonia, and ammonia can be stored at a relatively high density as metal amines (such as $Mg(NH_3)_6Cl_2$).

Up to volumes of $5000\,m^3$ NH_3, the common technology to store NH_3 is as a liquid, pressurized at ambient temperature. The minimum required pressure is depending on the ambient temperature, but a typical value is about 10 bar (a). For larger volumes, the common way for storing NH_3 is as a liquid at ambient pressure and at the saturation temperature of about −33°C. To ensure containment, a double-wall tank system is applied. Boil-off NH_3 is captured and returned into the tanks by a redundant system.

A disadvantage is that the ammonia release from the metal amine complex and the subsequent decomposition of NH_3 into N_2/H_2 in the ideal case requires approximately 31% of the energy content in the stored hydrogen.

FURTHER READING

[1] J. Amouroux, P. Siffert, J.P. Massué, S. Cavadias, B. Trujillo, K. Hashimoto, P. Rutberg, S. Dresvin, X. Wang, Carbon dioxide: a new material for energy storage, Prog. Nat. Sci.: Mater. Int. 24 (2014) 295–304.

[2] R. Carnegie, D. Gotham, D. Nderitu, P.V. Preckel, Utility Scale Energy Storage Systems, Benefits, Applications, and Technologies, State Utility Forecasting Group, 2013.

[3] H. Chen, T.N. Cong, W. Yang, C. Tan, Y. Li, Y. Ding, Progress in electrical energy storage system: a critical review, Prog. Nat. Sci. 19 (2009) 291–312.

[4] H. Chen, T.N. Cong, W. Yang, C. Tan, Y. Li, Y. Ding, Progress in electrical energy storage system: a critical review, Prog. Nat. Sci. 19 (2009) 291–312.

[5] J.M. Christensen, P.V. Hendriksen, J.-D. Grunwaldt, A.D. Jensen, Chemical energy storage, in: I.H.H. Larsen, L.S. Petersen (Eds.), Energy Storage Options for Future Sustainable Energy Systems, Technical University of Denmark, 2013, pp. 47–52. DTU International Energy Report.
[6] F. Díaz-González, A. Sumper, O. Gomis-Bellmunt, R. Villafáfila-Robles, A review of energy storage technologies for wind power applications, Renew. Sust. Energ. Rev. 16 (2012) 2154–2171.
[7] D.H. Doughty, P.C. Butler, A.A. Akhil, N.H. Clark, J.D. Boyes, Batteries for large-scale stationary electrical energy storage, Electrochem. Soc. Interface 19 (2010) 49–53.
[8] Encyclopaedia Britannica, Battery: Nickel-Cadmium Battery, Available at: http://kids.britannica.com/comptons/art-52969/A-cutaway-diagram-shows-anickelcadmium-rechargeable-cell-Its, April 20, 2018.
[9] EPRI, The Future Role for Energy Storage in the UK Main Report, Energy Research Partnership (ERP) technology report published, June 2011.
[10] IEC Market Strategy Board, Electrical Energy Storage, White Paper, International Electrotechnical Commission IEC WP EES:2011–12(en), 2012.
[11] A.M.F. Inamuddin, A.M. Asiri, S. Zaidi (Eds.), Electrochemical Capacitors-Theory, Materials and Applications, vol. 26, Materials Research Foundations, ISBN 978-1-945291-56-2, 2018.
[12] B.K. Kim, S. Sy, A. Yu, J. Zhang, Electrochemical supercapacitors for energy storage and conversion, in: Handbook of Clean Energy Systems, John Wiley & Sons Ltd., 2015, pp. 1–25
[13] P. Leung, X. Li, C.P. de León, L. Berlouis, C.T.J. Low, F.C. Walsh, Progress in redox flow batteries, remaining challenges and their applications in energy storage, RSC Adv. 2 (2012) 10125–10156.
[14] X. Lu, G. Xia, J.P. Lemmon, Z. Yang, Advanced materials for sodium-beta alumina batteries: status, challenges and perspectives, J. Power Sources 195 (2010) 2431–2442.
[15] A.S. Muñoz, M. Garcia, M. Gerlich, in: Sensible—deliverable: overview of storage technologies, European Union's Horizon 2020 Research and Innovation Programme Under Grant Agreement No. 645963, 2016.
[16] A.M. Namisnyk, A Survey of Electrochemical Supercapacitor Technology, BS Project Report, University of Technology, Sydney, 2003.
[17] R.H. Newnham, W.G.A. Baldsing, A.F. Hollenkamp, O.V. Lim, C.G. Phyland, D.A. J. Rand, J.M. Rosali, D.J. Vella, L.H. Vu, Advancement of Valve-Regulated Lead-Acid Battery Technology for Hybrid-Electric and Electric Vehicle, Advanced Lead-Acid Battery Consortium, Durham, NC, 2002.
[18] OECD/IEA, in: Technology roadmap energy storage, 2014 International Energy Agency 9 rue de la Fédération 75739 Paris Cedex 15, France, 2014.
[19] S.T. Revankar, P. Majumdar, Fuel Cells: Principles, Design, and Analysis, CRC Press, 2014. ISBN 978-1-42-008968-4.
[20] Royal Society of Chemistry, Energy Storage Technologies, Available: http://www.rsc.org/Membership/Networking/InterestGroups/ESEF/storage/energystoragetechnologies.asp, April 20, 2018.
[21] R. Schlogl (Ed.), Chemical Energy Storage, Walter de Gruyter GmbH, Berlin/Boston, ISBN 978-3-11-026407-4, 2013.
[22] F. Schüth, Chemical compounds for energy storage-review, Chem. Ing. Tech. 83 (2011) 1984–1993.
[23] G. Strbac, M. Aunedi, D. Pudjianto, et al., Strategic assessment of the role and value of energy storage systems in the UK low carbon energy future, in: Report for Carbon Trust, Imperial College, London, April 23, 2018. Published June 2012. Available at: http://www.carbontrust.com/media/129310/energy-storage-systems-role-valuestrategic-.

[24] T.B. Atwater, A. Dobley, Metal/air batteries, in: Lindens Handbook of Batteries, McGraw-Hill, New York, ISBN: 978-0-07-1624190, 2011.
[25] R.S. Treptow, The lead–acid battery: its voltage in theory and in practice, J. Chem. Educ. 79 (2002) 334–338.
[26] UK DTI, Review of Electrical Energy Storage Technologies and Systems and of Their Potential for the UK, DTI Report. DG/DTI/00055/00/00, UK Department of Trade and Industry, 2004.
[27] Z. Yang, J. Zhang, M.C.W. Kintner-Meyer, X. Lu, D. Choi, J.P. Lemmon, J. Liu, Electrochemical Energy Storage for Green Grid, Pacific Northwest National Laboratory, Richland, WA, 2010.
[28] Z. Yang, J. Zhang, M.C.W. Kintner-Meyer, X. Lu, D. Choi, J.P. Lemmon, J. Liu, Electrochemical energy storage for green grid, Chem. Rev. 111 (2010) 3577–3613.

CHAPTER SEVEN

Packed Bed Thermal Storage for LWRs

Graham Wilson*, Sudhansu Sahoo†, Piyush Sabharwal‡, Hitesh Bindra*
*Department of Mechanical and Nuclear Engineering, Kansas State University, Manhattan, KS, United States
†College of Engineering, Bhubneshwar, Orissa, India
‡Idaho National Laboratory, Idaho Falls, ID, United States

7.1 THERMAL STORAGE FOR LWRs

The most widely implemented nuclear reactors throughout the world are light water-cooled reactors (LWRs), which produce saturated steam to operate steam turbines on the Rankine cycle principles. The two most common types of LWRs are boiling water reactors (BWRs), which directly produce saturated steam from the heat of the reactor core, and pressurized water reactors (PWRs). In a PWR, pressurized light water is circulated to remove reactor heat and transfer it to a secondary side via steam generators or intermediate heat exchangers. The thermal energy of the pressurized light water that is transferred to the secondary coolant (i.e., water) converts the water to steam for use in an outer Rankine cycle loop. Therefore, there are two possible routes to store thermal energy from LWRs—(a) extracting the thermal energy of the primary coolant into a prospective thermal energy storage (TES) system through the secondary heat transfer loop (only possible in case of PWRs); (b) extracting the energy of the steam directly or indirectly into TES media.

7.1.1 Thermal Storage Options and Integration Concepts

TES methods are the most economical among all proposed grid-scale energy storage solutions, and are naturally compatible with nuclear power plants (NPPs). There are various options to store thermal energy, which have been developed or are currently being researched. These options include the use of steam accumulators, sensible heat of a liquid or a solid, latent heat of fusion, and reversible thermochemical reactions. These methods can be

Storage and Hybridization of Nuclear Energy
https://doi.org/10.1016/B978-0-12-813975-2.00007-7

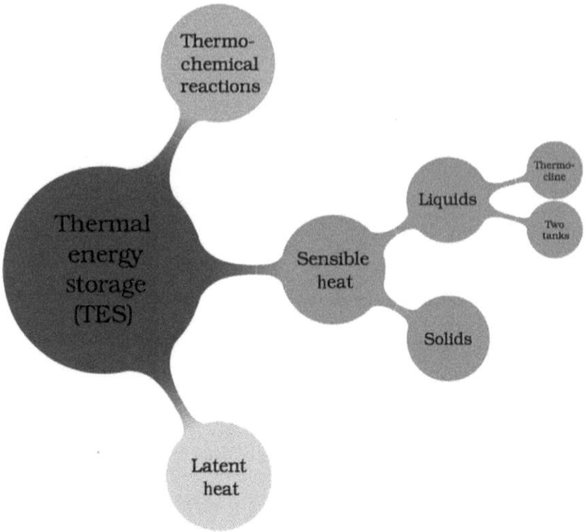

Fig. 7.1 Different thermal storage options which can be integrated with nuclear energy systems.

classified into three main categories: (i) sensible heat; (ii) latent heat; and (iii) thermochemical reactions (Fig. 7.1). All of these methods have their own relative advantages or disadvantages depending upon the requirements. All types of thermal storage systems have potential, but sensible heat storage solutions have reached a higher level of technological readiness. Any large or grid-scale integration requires more robust solutions, which implies that sensible heat storage options are preferred over other systems. It has been shown by Bindra et al. [1] that sensible heat storage has much higher exergy efficiency for high energy density storage design as compared to latent heat storage methods.

Sensible heat storage is obviously possible only in select liquids and solids, as the low thermal capacity of gases would render this method unfeasible. Currently, there are some thermal storage solutions such as molten nitrate salt, also known as solar salt (40%KNO_3 + 60%$NaNO_3$), which present very low technological risk and a high deployment potential. These solutions can be good candidates for molten salt or gas cooled high-temperature reactors [2], but have some limitations for integration into existing NPPs.

7.1.2 Liquid vs. Solid Sensible Heat Storage

Both solid and liquid sensible energy storage solutions have unique advantages for large-scale energy storage. Examples of liquid storage solutions for large-scale power plant applications include molten salts (nitrate), liquid metals, pressurized water, and heating oils (Therminol, Dowtherm, etc.). Liquid storage solutions also serve as a heat transfer fluid (HTF), which makes the system easier to design and control. Two-tank TES systems with one tank for hot liquid and one tank for cold liquid have been designed and deployed for large-scale concentrating solar power applications [3]. Integration of solid sensible energy storage solutions is more complex as it involves an additional HTF to deliver the energy into storage media and recover energy from the solids. There are two different classes of solid storage solutions, which have been looked at in the past—firebricks and metals. Firebricks constitute ceramics such as alumina, magnesite, silica, clay, etc., and metals include materials such as cast iron, steel, etc.

Energy density of sensible heat storage systems is dependent upon material density, specific heat, and temperature differential. Due to these reasons molten salt is the most preferred choice among liquids as it can be heated to high temperatures (570°C) without any degradation or pressurization requirements. Molten salts are also more economical from a large-scale production standpoint, but their applications are limited to high-temperature systems due to their higher freezing point (220°C). Liquids such as Therminol or Dowtherm can overcome those limitations by providing a wider temperature operating range [4]. In the case of solids, firebricks or refractory materials have an advantage over metals because of the cost; however, metals have a much higher energy density, which makes them less bulky for storing the same amount of energy. Upper and lower temperature constraints for solids are more flexible than liquids while providing substantially higher energy density. Liquids provide more flexibility in process design as they can be used as the HTF, whereas solid thermal storage mostly uses fluid media to store or recover energy.

The selection of a particular material as a sensible heat thermal storage option is largely dependent upon the operating and process conditions of the nuclear heat source and reactor coolant. Existing NPPs operate at lower temperatures than advanced reactors and only use pressurized water as the main coolant, whereas advanced reactors may use molten salts or high-temperature gases as the main coolant. Storing thermal energy in technologically mature liquid or solid materials can be achieved if systems are designed with a better understanding of thermal transport processes during integration.

7.1.3 Limitations of Established Process Solutions

Due to material and reactor safety limitations, most LWRs produce steam at 280°C or less. At these temperatures, thermodynamic efficiency is close to 35%, which is considerably lower than many fossil-fueled thermal power plants. Therefore, one of the prime requirements is to avoid any further reduction in efficiency due to storage integration. In PWRs, the losses due to heat transfer between the pressurized water and steam is one of the significant reasons for exergy destruction. The ideal configuration for storage integration is to store the energy of the primary reactor coolant (RC), but the coolant is considered as one of the intermediary layers for radioactivity containment, so for safety measures the coolant is generally not allowed to leave the containment building. This safety philosophy and large volume requirements for TES systems can only be met if the heat exchanger, which exchanges heat between RC and HTF, is housed inside the reactor containment building. The hot HTF is brought outside the reactor containment building to store energy in a TES system. As described earlier, Dowtherm and Therminol, or any other synthetic HTF, which remains liquid from room temperature to the highest temperature of RC, can be used as an HTF and heat storage media. These liquid type sensible heat energy storage options can be integrated with a PWR type NPP using a two-tank storage configuration (Fig. 7.2), which has been successfully demonstrated for molten salts in solar thermal systems. Exergy analysis for integrating these liquid type TESs with LWRs and LW-SMRs shows that the technical feasibility index of integrating TES is very high [4]. But in order to accommodate additional heat exchange equipment considerable changes in

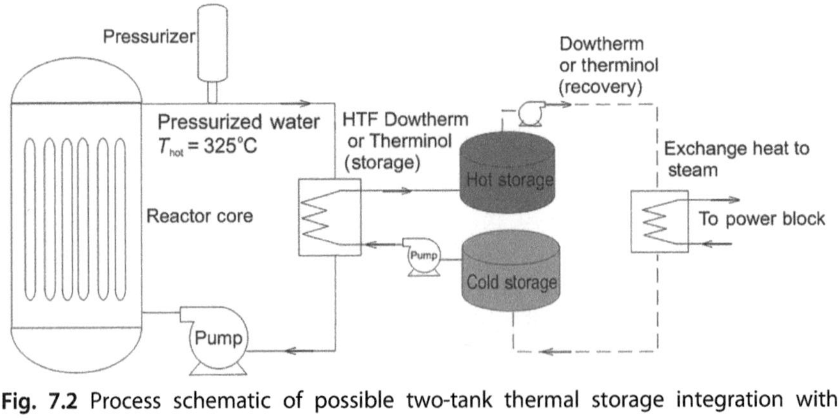

Fig. 7.2 Process schematic of possible two-tank thermal storage integration with a PWR.

the layout of the reactor building will be needed, which can also create stringent regulatory requirements. In addition, most of the LW-SMR-based NPP designs cannot accommodate additional heat exchange equipment, where the volume of the reactor containment for a standard 1000 MWe NPP is much greater than a NuScale 50 MWe LW-SMR (http://www.nuscalepower.com/).

Another alternative of storing heat energy in LWR plants is transferring the energy from steam produced in steam generators into an HTF and energy storage media. However, indirect transfer or exchange of latent heat is associated with the pinch point effect, resulting in inefficient heat transfer. Heat exchange between steam and another HTF during the storage process, and then heat recovery again via indirect heat exchange, will result in a considerable negative impact on the overall thermodynamic efficiency.

Another existing popular option is to use a steam accumulator, which stores dry or wet steam directly inside the pressure vessels; however, due to large volumetric requirements or sliding pressure it is not a viable option for several thousand megawatt-hours of grid-scale energy storage.

7.2 SOLID MEDIA FOR STORING ENERGY FROM LW-SMRs

A new method of integrating packed bed TES systems with LW-SMRs is presented here, where steam produced in the secondary side of LW-SMRs can be injected into the packed bed of inert particles with sufficient thermal conductivity and capacity to efficiently store energy with high energy density (schematic shown in Fig. 7.3). A typical packed bed TES system consists of a fluid path, a storage vessel internally lined with refractory or insulating material and a packing material. When electricity prices are low, steam is diverted from the turbines to a packed bed (rock or metal) in a pressure vessel where steam heats the pebbles while condensing. The high surface area of the pebble bed allows rapid steam condensation and thus very fast response with decreased steam to the nuclear turbine and decreased electricity to the grid. When prices are high, water is added to the pebble bed to produce high-pressure steam for electricity production. Also, stored heat can be recovered using any process fluid for any process heat application. During peak electricity demand, steam produced by the packed bed can augment steam from the reactor entering the turbine, delivering more electricity to the grid than the reactor could alone. While the reactor

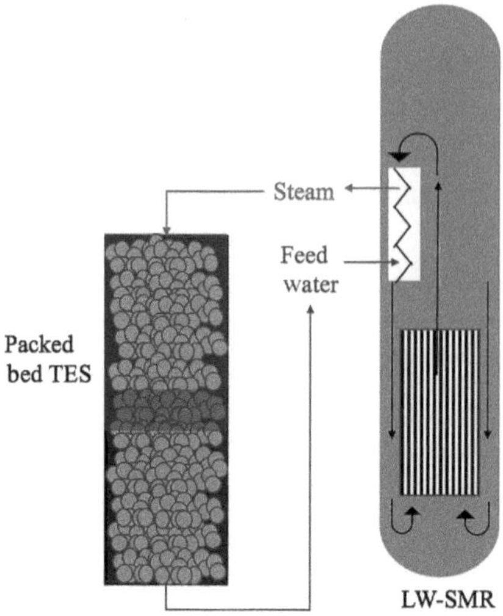

Fig. 7.3 A schematic showing steam flow from SMR into packed bed TES.

core may operate at constant power, the plant can rapidly change the amount of electrical power sent to the grid to meet market needs.

Packed bed type TES systems have been invented, designed, and optimized to work for single-phase HTFs [1, 4–9]. The design of packed bed type TES systems was motivated by geothermal reservoirs and high-temperature process industries.

7.2.1 History and Background: Sensible Heat Storage in Solids

Geothermal energy stored in the planet Earth is a common, naturally available example of sensible heat storage in solids. Although spatiotemporal distribution or release of this energy is highly random in nature, the energy density at certain locations on the Earth's surface is sufficient to run turbo-machinery or generate electricity. This natural energy storage reserve is the design motivation for primarily all types of solid-based TES systems. Key fundamental principles for the design of a storage system are: large reservoir volume, low surface area to volume ratios, high-temperature stability, high melting point, and chemical inertness to enclosures or HTFs. There are two types of solids, which have been explored for sensible energy storage—refractory materials and stable metals or their alloys. Clay, alumina, feolite, and magnesite in the form of firebricks or particles constitute refractory materials. Metals and their alloys include materials such as cast iron. Refractory materials are generally two times bulkier than iron alloys for the same amount of energy storage capacity but are more economical. Arrangement of solids for storing heat can be in the form of randomly packed or ordered configurations. Most of the solid sensible energy storage designs need HTFs to store or recover energy from solids, except for a few recent designs. The design layout of large-scale solid media thermal storage systems originated from industrial regenerative heat exchangers and catalytic bed reactors widely used in the process and energy industry.

7.2.2 Regenerative Counter-Current Heat Exchangers

Counter-current regenerative heat exchangers are preferred over other standard shell and tube or plate-type heat exchangers for high efficiency indirect heat exchange between two gaseous fluid streams. Counter-current thermal regenerative heat exchangers operate by temporarily storing the heat of a hot fluid in the solid permeable packing media and then releasing this heat into a colder fluid stream. While the surfaces of the packed solids are exposed to hot fluid, the sensible heat is getting stored. After the end of the storage phase, a reverse flow of cold fluid is initiated, and the stream is sent through the same flow geometry to recover the stored sensible heat. The switching of the bed from the hot to the cold fluid stream is obtained with the help of a set of isolation valves. The structure of the storage media in these heat exchangers is of the checker work form in which a matrix of solids is casted into regular and repeating periodic cells with fluid paths in between these solid structures. Two types of system configurations have been used in most of the industrial regenerators—Fixed bed type and Rotary type [10]. Fixed bed type regenerators were invented for use in blast furnaces and were named after their inventor as "Cowper stoves" (Fig. 7.4). Rotary type regenerators, or "Ljungstrom," are also the namesake of their creator. Current design of solid sensible heat storage systems is directly inspired by the design and operation of the "Cowper Stove," except for the actual storage media configuration, sizing, and material selection.

7.2.3 Catalytic Beds and Reverse-Flow

Randomly packed solid particles of different shape, size, and materials have seen a wide range of applications in chemical industry, particularly for heat transfer, mass transfer, and chemical reaction engineering. In the case of high-temperature catalytic reactors, a packed bed of ceramic particles enables virtually uniform heat removal from the bed by allowing nearly plug flow of fluid media. Due to heat generation inside the catalytic reactors,

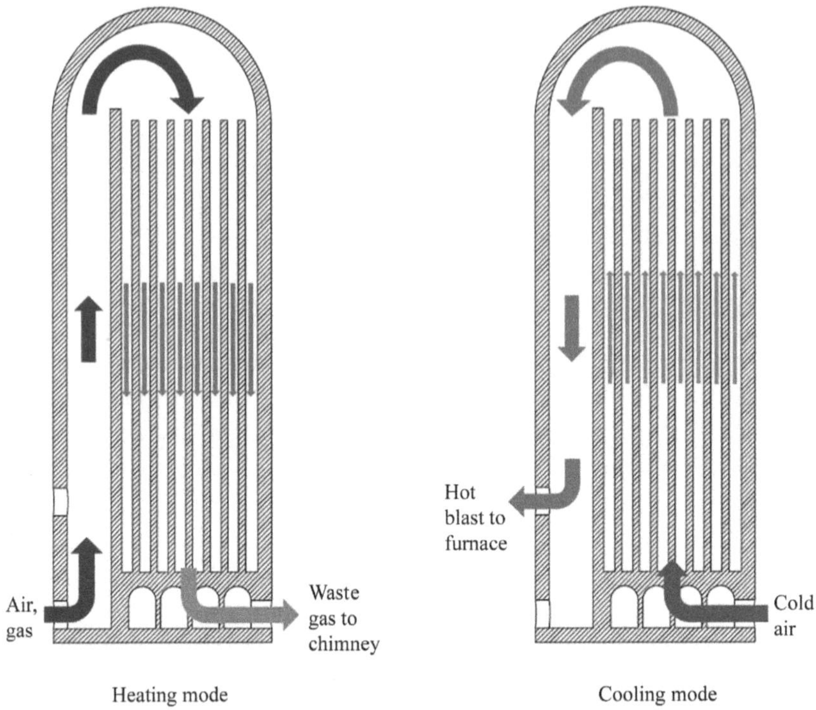

Fig. 7.4 Counter-current heat exchange operation in a Cowper Stove.

there have been several optimization studies on fixed or fluidized bed operation of packed beds. Reverse-flow operation in a fixed bed catalytic reactor, which was described by Matros et al. [11], is an example of process improvement under unsteady-state conditions where the catalyst quickly adjusts to changes in temperature and reaction mixture composition. There are different ways to achieve unsteady or periodic variations in the local or global reactor behavior. The basic mode of operation is very simple and involves a periodic reversal of cold gas flow to a preheated fixed catalyst bed. Reverse-flow catalytic fixed bed reactors have been preferred over steady-state traditional stationary reactors for maintaining high activity in sustaining exothermic-endothermic reversible chemical reactions while recovering the heat for efficient secondary use. These heterogeneous reverse-flow catalytic reactors show better efficiency due to two major factors: (1) the steep and big change in the local catalytic bed temperature observed during the reverse flow injection of cold fluid into an initially hot bed; and (2) the high relative thermal capacity of the catalyst as compared to the fluid, leading to suitable local and global reactor temperatures in the transient operation mode.

These design principles, which are prime contributors to the high performance of fixed bed catalytic reactors, provide the design concept for high-temperature packed bed type thermal storage.

7.3 THERMOCLINES AND STRATIFICATION

The primitive designs of sensible heat storage systems were based upon first law considerations, which emphasized higher energy density and efficiency. But as most examples of grid-scale energy storage are associated with work done, second law-based analysis is more suitable. The subjects of the first and second law analyses are the recovery of energy from a heated system and the extraction of useful work from such a system, respectively. The measure of this useful work is defined as exergy. It is also defined as potential energy of the system (with temperature T_s) before it reaches equilibrium with its surroundings (with temperature T_0). In mathematical form, exergy per unit heat capacity can be written as (see description of symbols in the "Nomenclature" section at the end of this chapter)

$$\Theta = (T_s - T_0) - T_0 \log(T_s/T_0) \tag{7.1}$$

In a constant temperature or constant volume system, exergy per unit heat capacity increases monotonically with increasing temperature for a system temperature always greater than the surrounding temperature.

Bejan [12] and Krane [13] showed the need of second law or exergy analysis for bath-type systems undergoing thermal storage. Krane's [13] analysis showed that under perfect mixing inside a storage system the exergy efficiency for a storage-recovery cycle is less than 27%. These analyses assumed that the storage system is a well-mixed liquid bath and modeled it as a lumped system. In the case of a well-mixed system during the heat storing process, the inlet stream temperature is always higher than that of the outlet stream, which is in equilibrium with the bath temperature. Since the outlet stream temperature is higher than the initial system temperature, some energy is lost as heat is stored. Similarly, during the recovery process the maximum temperature that can be obtained from a mixed tank is always less than the system temperature at the beginning of the process. The analyses [12, 13] showed that the exit stream temperature recovered from the stored heat is significantly lower than that of the inlet stream during heat storage, resulting in a lower Carnot efficiency for the system. Due to the lumped capacitance assumption, this model clearly underestimates the exergy efficiency of a storage-recovery process for any system. A complete opposite of a well-mixed system is a system in which the hot energy stream is introduced into the storage, pushing out a lower-temperature ("cold") stream without any mixing [13] between the two streams. In other words, a perfect plug flow system is expected to deliver high exergy efficiency.

7.3.1 Mixed vs Plug Flow

TES systems should be able to store energy at the required temperature in order to achieve the desired work output or exergy. The ideal storage system stores and recovers energy at the same temperature (i.e., 100% exergy recovery). Under the mixed flow condition as described earlier, two mean temperatures are defined, one based on energy segmentation (T_m) and another based on exergy segmentation (T_{ex}). The temperature of the storage media when perfectly mixed is equal to

$$T_m = \frac{1}{H}\int_0^H T_p dp \tag{7.2}$$

which implies that the whole system response is lumped and that the corresponding exergy per unit heat capacity of such a system is given by

$$\Theta = (T_m - T_0) - T_0 \log(T_m/T_0) \tag{7.3}$$

Whereas, in the case of a system completely stratified [1, 14] into the hot and cold zones expected under plug flow conditions, the exergy per unit heat capacity of each stratified section is

$$\Theta_p = (T_p - T_0) - T_0 \log(T_p/T_0) \tag{7.4}$$

and thus, total exergy per unit heat capacity of the system in this stratified mode can be expressed as

$$\Theta_S = \frac{1}{H}\int_0^H \Theta_p dp \tag{7.5}$$

The difference of exergy per unit heat capacity between plug mode and mixed mode can then be simplified to

$$\Delta\Theta = \Theta_S - \Theta = \frac{1}{H}\int_0^H \Theta_p dp - \Theta \tag{7.6}$$

$$\Delta\Theta = \Theta_S - \Theta = -T_0\left(\frac{1}{H}\int_0^H \log T dp - \log \frac{1}{H}\int_0^H T dp\right) \tag{7.7}$$

Rewriting the integrals in their Riemannian form results in

$$\Delta\Theta = \Theta_S - \Theta = -T_0 \lim_{\Delta l \to 0} \left(\sum_i \log[T_i]\Delta l/H - \log[(\sum_i T_i \Delta l)/H]\right) \tag{7.8}$$

Substituting $H = N\Delta l$ leads to

$$\Delta\Theta = \Theta_S - \Theta = T_0\left(\log[\sum_i T_i/N] - \log[\prod_i T_i]^{1/N}\right) \tag{7.9}$$

As the arithmetic mean $\sum T_i/N$ is always greater than the geometric mean $\prod T_i^{1/N}$, the quantity $\Delta\Theta > 0$. In other words, thermally stratified systems have higher exergy than well-mixed systems (Fig. 7.5). Several schemes involving stratification in liquid tanks and packed beds have been developed to attempt to reduce exergy loss but packed bed systems, unlike liquid tank systems, lend themselves to feasible designs.

7.3.2 Steam Injection in Packed Beds

With motivations described earlier, it is now imperative to evaluate the thermal performance of packed bed type storage solutions before considering them for practical use. Unfortunately, there have been very limited experimental or theoretical studies, which can help in modeling the thermal behavior of a packed bed upon steam injection. There is no existing report in the literature describing a mathematical model to predict the thermal behavior of such a system. A basic energy balance model is presented in this section to evaluate the thermal behavior of the bed during a storage or recovery process.

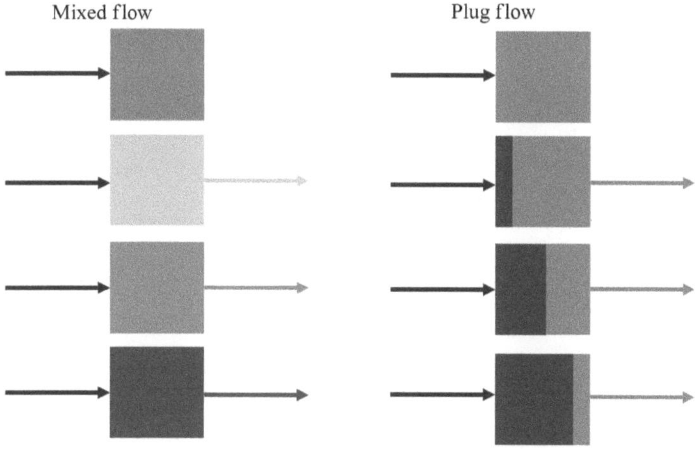

Fig. 7.5 Qualitative behavior depicting plug versus mixed flow type thermal storage.

There are some existing popular packed bed TES models, which consider only the sensible energy of HTFs [1,4,5]; however, these models lack phase change behavior in the fluid stream, which is essentially important for predicting this integration. A theoretical evaluation of the heat storage and recovery process that employs steam or water as the HTF and a bed of ceramic particles as the storage media is accomplished through the development of a three-phase (solid-liquid-vapor) mathematical model. The model is based on a homogeneous mixture of two phases during a phase change (boiling/condensation). The remaining part of the model is exactly same as the two-equation model used for the single-phase fluid/solid heat transfer model described extensively in the literature [1,5,15]. The basic assumptions for this model are—(a) There are no pressure changes during phase changes. (b) There is no temperature gradient inside the particles; this is valid for particles with a small size and a high thermal conductivity (Biot number < 0.1). Lower thermal conductivities, higher velocities, and larger particle sizes can lead to a violation of this assumption. This means that the temperature inside the particles cannot be considered uniform, which is not suitable for a storage system as it causes spreading of the thermal front. (c) No dispersion occurs in the liquid or vapor phase due to thermal conduction. This assumption is relaxed here to simplify the model but will be discussed in detail later. (d) Latent heat can be approximated with heat capacity over a temperature differential of 1°C. As the δT is increased, the accuracy of the model is lowered, but the numerical complexity is decreased. (e) During phase change, the liquid-vapor phase forms a homogeneous mixture with no liquid-vapor stratification.

The model only considers energy and mass conservation equations for the fluid phase as shown by Eqs. (7.10), (7.11), respectively.

$$\frac{\partial \rho}{\partial t} + \frac{\partial \rho v}{\partial x} = 0 \tag{7.10}$$

$$\frac{\partial \epsilon \rho h}{\partial t} + \frac{\partial \epsilon \rho h v}{\partial x} = -h_v (T - T_s) \tag{7.11}$$

However, as the solid media mass remains constant, only energy conservation is modeled for solid media.

$$(1 - \epsilon) \rho_s c_{ps} \frac{\partial T_s}{\partial t} = h_v (T - T_s) \tag{7.12}$$

Solid enthalpy can be related to the reference temperature directly, whereas enthalpy of the fluid phase can be expressed as

$$H = C_{pl}(T_{sat} - T_0) + f\lambda_{lg} + C_{pg}(T - T_{sat}) \tag{7.13}$$

The phase change term $f\lambda_{lg}$ can be converted to a simplified sensible heat approximation (sharp temperature jump around phase change temperature).

$$H = C_{pl}(T^{I}_{sat} - T_0) + C_{p,jump}(T^{II}_{sat} - T^{I}_{sat}) + C_{pg}(T - T^{II}_{sat}) \tag{7.14}$$

Density of the fluid phase can be described as,

$$\rho_f = \rho_g + (1 - f)\rho_{lg} \tag{7.15}$$

which can be approximated to

$$\rho_f = \rho_g + \left(1 - \left(\frac{T - T^{II}_{sat}}{T^{II}_{sat} - T^{I}_{sat}}\right)\right)\rho_{lg} \text{ if } T^{II}_{sat} > T > T^{I}_{sat} \tag{7.16}$$

The enthalpy jump approximation can be explicitly stated as,

$$C_{p,jump}(T^{II}_{sat} - T^{I}_{sat}) = (f^{II} - f^{I})\lambda_{lg} \tag{7.17}$$

$$f^{II} = 1 \;\; f^{I} = 0 \tag{7.18}$$

Due to the high rate of latent heat of enthalpy injection and the high surface heat transfer of packed beds, the temperature at any point in the bed reaches the peak saturation temperature as soon as the steam front reaches it [16], as shown in Fig. 7.6. This type of behavior is distinctly different than when a noncondensible hot gas is injected into the packed bed, where the bed never

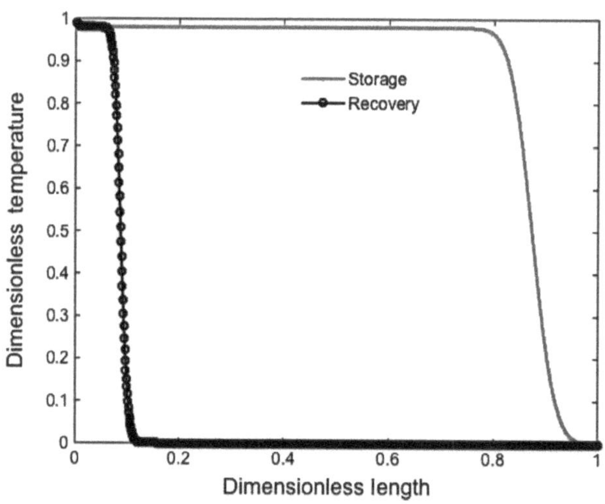

Fig. 7.6 Storage and recovery cycle calculations for a packed bed system with saturated conditions at 50 bars with alumina particles and steam as HTF.

reaches the peak saturation temperature [1] due to a higher heat loss rate. These results show that this storage process is economically feasible because the bed utilization factor is close to 75% of the bed volume. Higher utilization implies higher energy density and in-turn lower capital costs. During saturated steam injection, condensate hold-up can significantly affect the dispersion of the thermal front; thus, understanding and modeling the behavior of the overall system remains a complex process.

7.3.3 Effective Thermal Diffusivity: Impact on Axial Temperature Front

Thermal stratification or the division of a volume into strata of different local thermal conditions can be achieved by an ideal plug flow scenario. Practically, however, thermal front propagation is always associated with diffusion, conduction, and local eddies which promote mixing at length scales much smaller than the overall system. It has been observed in the case of packed beds that it is much easier to control temperature in the plane normal to the flow direction. This is mostly achieved by realizing a uniform flow distribution at the packed bed inlet. But molecular or eddy thermal diffusivity is much more difficult to control in the axial, or flow, direction. Due to these reasons steep thermoclines are much harder to achieve and can result in a considerable energy or exergy loss along the exit flow stream (Fig. 7.7). A direct negative impact of such a design is that significant losses are associated with bringing the entire bed to the top temperature. This is also referred to as axial thermal dispersion, which is very difficult to model in

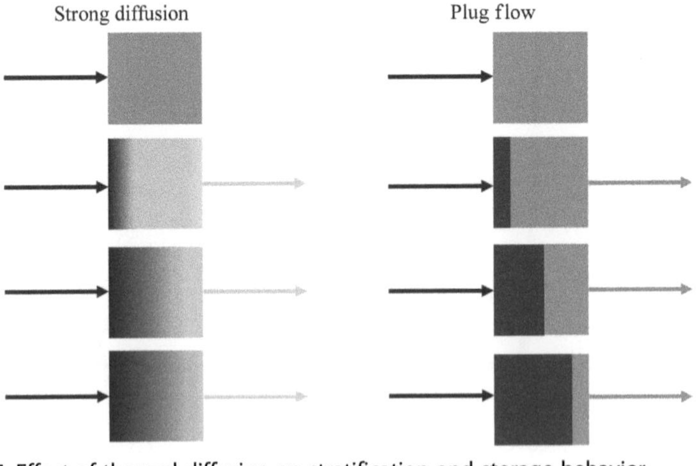

Fig. 7.7 Effect of thermal diffusion on stratification and storage behavior.

packed beds due to complex fluid dynamics and the resulting eddy thermal diffusivity around randomly packed particles. In the case of phase change in the fluid stream, the predictive modeling of axial dispersion is even more complex and far less understood.

Axial dispersion is an important parameter used to design packed bed TES systems; therefore, its accurate measurement or prediction can dramatically improve the confidence level in the technology. Recent experimental studies were conducted with a packed bed of alumina particles and saturated steam at atmospheric pressure as the HTF. A Rayleigh backscattering-based distributed temperature sensing fiber optics system was deployed in experimental and pilot scale geometries to obtain more precise measurements of axial dispersion than thermocouple arrangements can. Significant findings of these experiments are briefly discussed in the following paragraphs (for more details refer to recent publications [15,17,18]).

7.3.3.1 Slow Steam Injection

Upon injection of steam into the bed, there are two possible thermal transport mechanisms—advection and conduction modes at different spatial locations and different time frames. Near the entry port where steam is introduced, in this present experimental set-up from the top, the temperature of the bed and the fluid stream becomes almost equal to the steam inlet, or saturation, temperature within a very short time interval. With the steam supply continuously available, irrespective of injection rate, the bed temperature at the top is always maintained at a constant top temperature (i.e., the steam saturation temperature). This constant bed temperature at the top will conduct heat from the top to the bottom of the bed due to the nonnegligible thermal conductivity of the alumina particles and water condensate in the bed (i.e., the conduction mechanism). Simultaneously, the steam injected into the bed carries some amount of energy as it moves in the bed (i.e., the advection mechanism).

Due to the slow injection rate, the initial rise in temperature at more axially distant locations will be dominated by the conduction mechanism. As the steam or two-phase mixture front, which is at a temperature near the saturation temperature, reaches those regions located far away from the injection point there is a sudden change in temperature. This effect can be seen in Fig. 7.8 based on the temperature profiles obtained at earlier time stamps. The rate of increase of temperature for different axial locations is divided into two distinct regimes with two distinct slopes. The domain initially near the inlet, which has a higher temperature slope, is governed by the

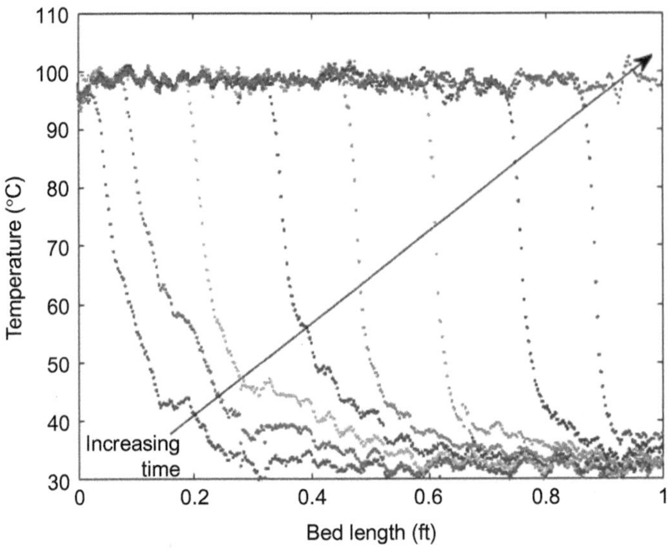

Fig. 7.8 Axial thermal response data from ODISI-B inside an alumina packed bed upon slow steam injection (1 lb/h steam at atmospheric pressure).

advection mechanism, while further downstream regimes with a lower slope are initially governed by conduction. Similar effects were quantitatively predicted and experimentally observed by Woods et al. in liquid-vapor flows around porous beds [19].

7.3.3.2 Fast Steam Injection

Based on the explanations provided in the previous section, it is expected that the advection term will be higher in fast injection experiments than in slow injection experiments. The higher advection term implies that the total amount of influx enthalpy carried by the steam or two-phase mixture is much higher; thus, as the fluid stream moves through the bed it is equilibrating the bed to the saturation temperature at an almost constant rate at all spatial locations. Due to the much higher rate of enthalpy injection in the bed because of the dominant advection term, the effects of conduction will not be observable during the time it takes to completely saturate the bed to the top temperature. This phenomenon of a smaller temperature dispersion due to conduction can be seen in the experimental results for fast injection in Fig. 7.9. This is also due to small length (1 ft) of the test setup used for these studies.

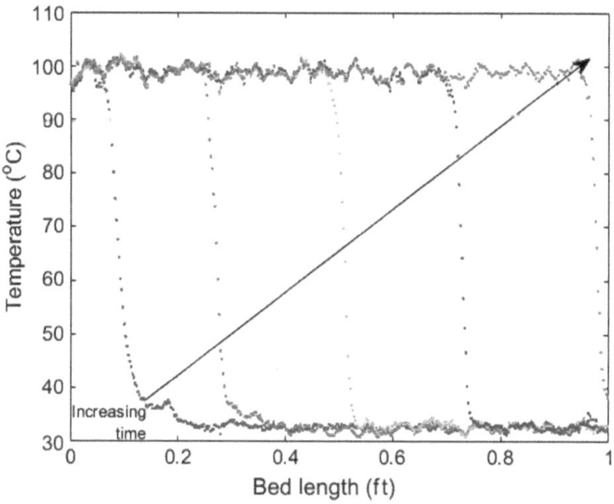

Fig. 7.9 Thermal response for fast steam injection (4.5 lbs/h).

Compared to the slow injection case, the time taken for the bed to reach peak temperature throughout in the fast injection case is shorter.

It can be concluded that during steam injection, under all flow rates, the temperature front remains considerably steep (i.e., axial dispersion is minimal), as a function of time and position.

7.4 STATE OF THE ART AND FUTURE WORK

Thermal stratification with steep temperature gradients plays a critical role in defining the performance of packed bed TES systems. Temperature gradients are much sharper in the alumina packed bed upon steam injection as compared to any other experimental data on thermoclines available in the literature for any type of thermal storage system. After several hot-cold or steam-water injection test cycles at 300 psi with alumina [20], it has been confirmed that alumina rocks of 3–6 mm particle size do not exhibit any ratcheting under thermal cycles. So, the future potential of this method is highly promising for integrating heat storage with steam producing power plants including all types of LWRs. Experimental studies show that even with low-grade less expensive materials such as pea-gravel steep thermoclines can be obtained, but there is a need to characterize and evaluate the durability this low-cost material. Although the cost of pea gravel is much lower than the cost of alumina particles, the particles have a higher energy density. Other advantages of alumina particles are their higher thermal capacity and higher thermal conductivity. The energy density of a packed bed TES with granite rocks (the major constituent in the tested pea gravel) in the temperature range of existing NPPs is ~2 kWh (e)/ft^3. With the pea gravel as the storage medium, the cost with vessel, materials, piping, and auxiliaries is expected to be $30/kWh(e) for LWR operating range.

Future work on the deployment of these storage systems or to evaluate their exact economic potential needs to include the effect of higher pressure on the axial dispersion. For effective storage demonstration tests, a recovery cycle is also required at pilot scale, which will be another critical task for the future.

NOMENCLATURE

Variables	Description
T	Temperature
x	Distance
t	Time
v	Velocity
H	Enthalpy
h	Specific enthalpy
Θ	Exergy per unit heat capacity
ϵ	Porosity
f	Vapor phase mass fraction
λ	Latent heat
ρ	Density
h_v	Volumetric heat transfer coefficient
C_p	Heat capacity
c_p	Specific heat capacity

Subscript	Description
p	Segment
s	Solid or system
l	Liquid
g	Vapor
sat	Saturation
0	Reference
m	Mean
S	Stratified

ACKNOWLEDGEMENT

"This material is based upon work supported by the Department of Energy under Award Number(s) DE-NE0008498 and DE-AC07-05ID14517."

Disclaimer: "This report was prepared as an account of work sponsored by an agency of the United States Government. Neither the United States Government nor any agency thereof, nor any of their employees, makes any warranty, express or implied, or assumes any legal liability or responsibility for the accuracy, completeness, or usefulness of any information, apparatus, product, or process disclosed, or represents that its use would not infringe privately owned rights. Reference herein to any specific commercial product, process, or service by trade name, trademark, manufacturer, or otherwise does not necessarily constitute or imply its endorsement, recommendation, or favoring by the United States Government or any agency thereof. The views and opinions of authors expressed herein do not necessarily state or reflect those of the United States Government or any agency thereof."

REFERENCES

[1] H. Bindra, P. Bueno, J.F. Morris, R. Shinnar, Thermal analysis and exergy evaluation of packed bed thermal storage systems, Appl. Therm. Eng. 52 (2013) 255–263.
[2] C. Forsberg, D. Curtis, Meeting the needs of a nuclear-renewable electrical grid with a fluoride-salt-cooled high-temperature reactor coupled to a nuclear air-Brayton combined cycle power system, Nucl. Technol. 185 (3) (2014) 281–295.
[3] U. Hermann, et al., Two-tank molten salt storage for parabolic trough solar thermal plants, NREL Rep. 2011.
[4] J. Edwards, H. Bindra, P. Sabharwall, Exergy analysis of thermal energy storage options with nuclear power plants, Ann. Nucl. Energy 96 (2016) 104–111.
[5] H. Bindra, P. Bueno, J.F. Morris, Sliding flow method for exergetically efficient packed bed thermal storage, Appl. Therm. Eng. 54 (2014) 201–208.
[6] H. Bindra, R. Shinnar, Thermal energy storage for combined cycle power plants, (2017). U.S. Patent No. 9,540,957.
[7] H. Bindra, P. Bueno, Optimum process design of packed bed type thermal storage systems and other applications, (2016). U.S. Patent No. 9,475,023.
[8] M. Kawaji, H. Bindra, Methods for meeting localized peak loads in buildings and urban centers, U.S. Patent Application No. 15/029,953.
[9] H. Bindra, J. Edwards, D. Gould, U.S. Patent Application No. 62/339,576 (International Patent Application No. PCT/US2017/033566).
[10] F.W. Schmidt, A.J. Willmott, Thermal Energy Storage and Regeneration, Hemisphere Publishing Corporation, Washington, DC, 1981.
[11] Y.S. Matros, G.A. Bunimovich, Reverse-flow operation in fixed bed catalytic reactors, Catal. Rev. 38 (1) (1996) 1–68.
[12] A. Bejan, Two thermodynamic optima in the design of sensible heat units for energy storage, J. Heat Transfer. 100 (1978) 708–712.
[13] R.J. Krane, A second law analysis of the optimum design and operation of thermal energy storage systems, Int. J. Heat Mass Transfer 30 (1) (1987) 43–57.
[14] M.A. Rosen, The exergy of stratified thermal energy storages, Sol. Energy 71 (3) (2001) 173–185.
[15] J. Edwards, H. Bindra, P. Sabharwall, Steam condensation in packed bed, NURETH-16, Chicago IL, 2015. Paper No. 14054.

[16] D. Franken, J. Edwards, P. Sabharwall, H. Bindra, Synthetic heat transfer fluids as thermal energy storage media for existing NPPs, Transactions of American Nuclear Society, Annual Meeting, New Orleans, LA, 2016.
[17] J. Edwards, H. Bindra, An experimental study on storing thermal energy in packed beds with saturated steam as heat transfer fluid, Sol. Energy 157 (2017) 456–461.
[18] J.N. Edwards, Thermal Energy Storage for Nuclear Power Applications (Dissertation), Kansas State University, Dissertation, 2017.
[19] A. Woods, et al., The vaporization of a liquid front moving through a hot porous rock, J. Fluid Mech. 3443 (1997) 303–316.
[20] J.F. Morris, et al., A novel storage method for concentrating solar power plants allowing storage at high temperature, (2014). Final Report FG36-08GO18151.

CHAPTER EIGHT

Cryogenic Energy Storage and Its Integration With Nuclear Power Generation for Load Shift

Qinghua Yu*, Tongtong Zhang*, Xiaodong Peng*, Lin Cong*, Lige Tong†, Li Wang†, Xiaohui She*, Xiaosong Zhang‡, Xinjing Zhang§, Yongliang Li*, Haisheng Chen§, Yulong Ding*,†

*Birmingham Centre for Energy Storage, School of Chemical Engineering, University of Birmingham, Birmingham, United Kingdom
†School of Energy and Environmental Engineering, University of Science and Technology Beijing, Beijing, China
‡School of Energy and Environment, Southeast University, Nanjing, China
§Institute of Engineering Thermophysics, Chinese Academy of Sciences, Beijing, China

8.1 INTRODUCTION TO CRYOGENIC ENERGY STORAGE

8.1.1 Basic Principle and Technology Development History

Cryogenic energy storage (CES) refers to a technology that uses a cryogen such as liquid air or nitrogen as an energy storage medium [1]. Fig. 8.1 shows a schematic diagram of the technology. During off-peak hours, liquid air/nitrogen is produced in an air liquefaction plant and stored in cryogenic tanks at approximately atmospheric pressure (electric energy is stored). During peak hours, ambient heat is used to boil the cryogen to give a high-pressure gas, driving a turbine for electricity production (electric energy is discharged).

The air liquefaction process involves air compression, leading to the production of heat. The compression heat can be harvested and stored for use in the discharge process to superheat the air and increase the power output. The high-grade cold during the discharge process can be captured and stored for use in the charge process to reduce power consumption of air liquefaction. The effective use of the compression heat and high-grade cold can enhance the round-trip efficiency of the CES technology, which, for a large-scale stand-alone plant (>~10 MW), can reach ~60%. If additional heat sources (e.g., renewable heat and industrial waste heat) and cold sources

Storage and Hybridization of Nuclear Energy
https://doi.org/10.1016/B978-0-12-813975-2.00008-9

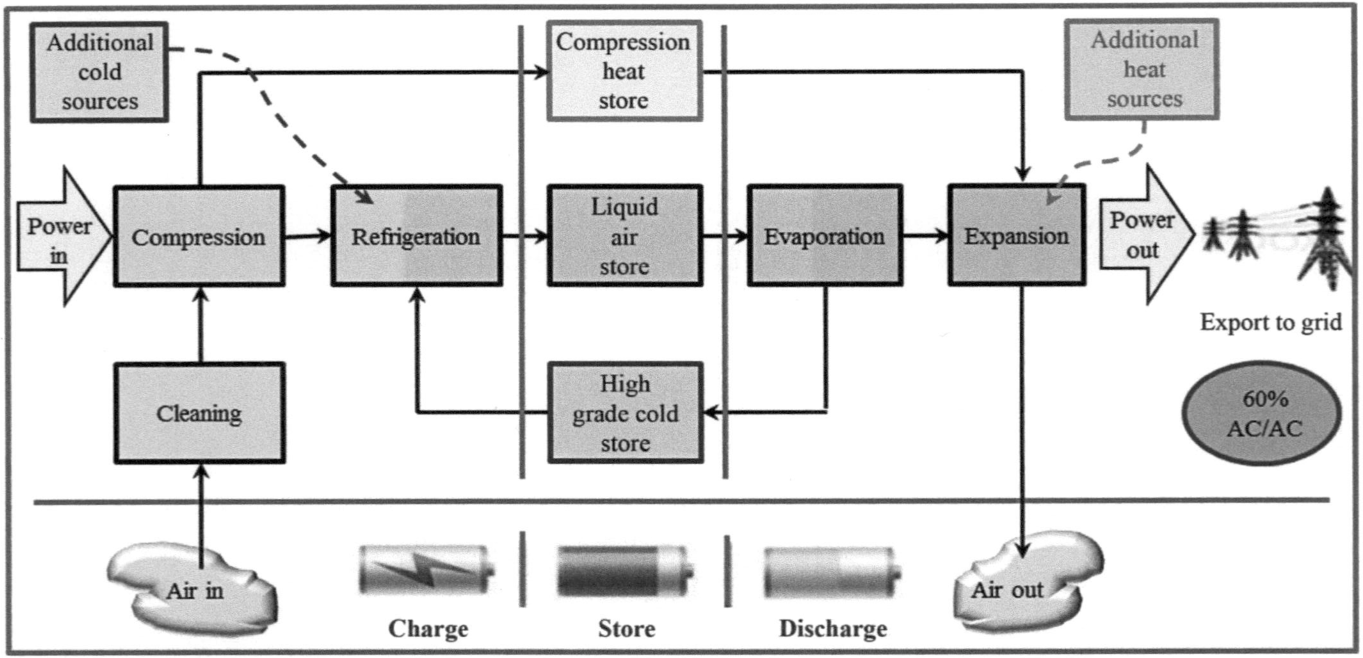

Fig. 8.1 The schematic diagram illustrating the principle of the CES technology.

(e.g., waste cold from the evaporation of liquefied natural gas) are available, a further enhancement in the round-trip efficiency could be achieved.

The use of cryogen as an energy storage medium can be dated back to 1899–1902 when cryogenic engines were first invented. The concept of the CES technology, however, was proposed much late in 1977 by researchers at the University of Newcastle upon Tyne in the United Kingdom for peak shaving of electricity grids [2]. Although the work at Newcastle involved mainly theoretical analyses, it led to subsequent developments of the technology by a number of industrial companies including Mitsubishi Heavy Industries [3] and Hitachi [4–6] of Japan, Expansion Energy [7], and Highview Power Storage [1, 8] and academic institutions particularly the Universities of Leeds and Birmingham (United Kingdom) [9–16].

The CES research over the past few decades can be divided into two categories. First, treating the CES technology as an extended alternative to compressed air energy storage (CAES) with the energy release process combined with conventional gas turbine technology, the work carried out by Mitsubishi Heavy Industries on a 2.6 MW pilot plant [3], Hitachi on an integrated system with a cold storage unit termed "regenerator" [4–6], and the University of Leeds on a CES integrated gas-fueled power plant with CO_2 capture [9, 12] falls into this category. Second, taking the CES technology as stand-alone storage facility without combustion process with waste-heat utilization in the discharge process, the world's first independent CES pilot plant (350 kW/2.5 MWh) built and tested between 2009 and 2012 by UK Highview Power Storage Ltd., based on the invention by researchers at the University of Leeds, falls into this category [1,17]. The pilot plant has now been relocated to the University of Birmingham (United Kingdom) for further research and development [14]. Current development of the CES technology includes further enhancement of the round-trip efficiency of the CES, for example, a recent research by the University of Birmingham suggesting an enhancement of the round-trip efficiency by ~9%–12% through effective utilization of excess heat of compression [15], and large-scale demonstration plants, for example, a 5 MW/15 MWh commercial demonstration plant built and tested by Highview Power Storage in collaboration with Viridor supported by the UK Department for Business, Energy and Industrial Strategy [18].

8.1.2 Process Diagram, Performance Evaluation and Application Range

Fig. 8.2 shows the process diagram of a typical CES plant [14], which, as described above, consists of three distinctive but interconnected and

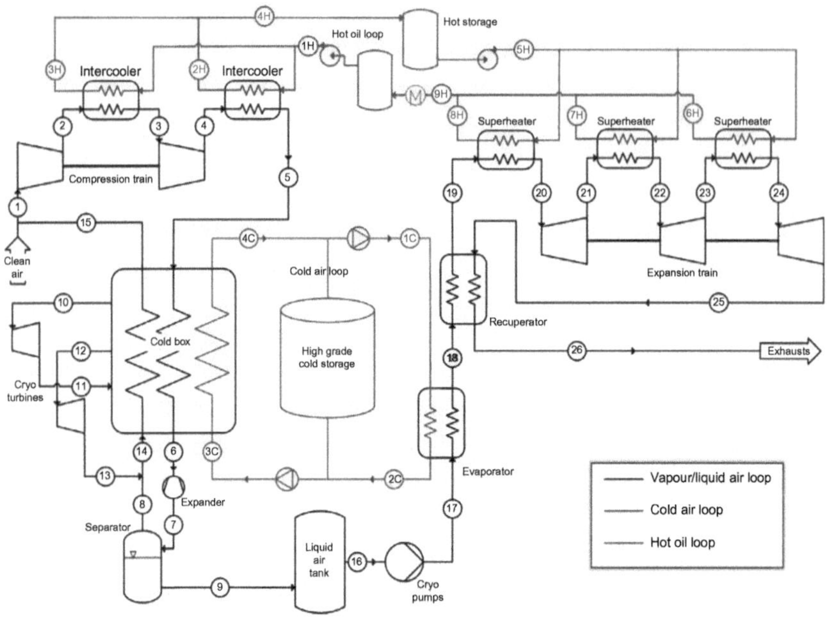

Fig. 8.2 Flow diagram of a typical CES plant (hot oil loop indicated by H and cold air loop indicated by C).

integrated subsystems: the air liquefaction unit, the storage units, and the power recovery unit. During charging, air from environment is liquefied through a serial of compression and expansion processes based on a modified Claude cycle. The compression process (streams 1–5) is equipped with two intercoolers, through which compression heat is stored. Here, the compression heat storage in the sensible form in a diathermic oil is considered as an example. Such an oil acts as both a heat-transfer fluid in the intercoolers and a storage medium in the hot storage tanks (streams 1H–5H). The high-pressure air from the last stage of compression cooled by the corresponding intercooler is further cooled in the cold box (streams 5–6) by the cold air (streams 14–15) from the gas/liquid separator and the cold air (streams 3C–4C) from the high-grade cold store (HGCS). The two air streams 10 and 12 in the cold box are expanded in two turbo expanders to improve the efficiency of the cooling process. Finally, the cooled air in the cold box expands in the cryoexpander, forming a mixture of gaseous and liquid air. The mixture is split in the separator with the liquid air stored in the cryogenic tank at about 80 K and close to the ambient pressure. During discharging, the stored liquid air is pressurized first by cryogenic pumps and then reheated in a series of heat exchangers including an evaporator, a recuperator, and a superheater by using the environmental heat and the compression heat. The cold energy released during liquid air vaporization (streams 17–18) is captured by a counter flow of heat-transfer fluid (streams 1C–2C) and stored in the HGCS. The HGCS can be realized using technologies such as packed beds of rocks as considered here and operated at nearly ambient pressure to reduce the costs and increase the safety of the HGCS. Finally, the heated high-pressure air expands in multistage turbines to generate electricity (streams 20–25).

A key thermodynamic performance indicator for the evaluation of the CES plant is the so-called round-trip efficiency, which is defined as the ratio of the work output in the energy release process to the power consumption in the energy storage process:

$$\eta_{RT} = \frac{W_{ER} \cdot t_{ER}}{W_{ES} \cdot t_{ES}} \tag{8.1}$$

where W_{ER} and W_{ES} are the power output in the discharge process and the power consumption in the charge process, respectively, and t_{ER} and t_{ES} are the discharge and charge durations, respectively. The power output in discharge process can be calculated by

$$W_{ER} = m_{16}((h_{20} - h_{21}) + (h_{22} - h_{23}) + (h_{24} - h_{25}) - (h_{17} - h_{16})) \tag{8.2}$$

where m_i is the air mass flow rate and h_i is the specific enthalpy with the subscript i representing state i as shown in Fig. 8.2. The power input during charge process is given by

$$W_{ES} = m_1((h_2 - h_1) + (h_4 - h_3) - (h_6 - h_7)) - m_{10}(h_{10} - h_{11}) \\ - m_{12}(h_{12} - h_{13}) \tag{8.3}$$

The durations of discharge and charge processes meet the following requirement:

$$\frac{m_{16} \cdot t_{ER}}{m_1 \cdot t_{ES}} = Y \tag{8.4}$$

where Y is the liquid air yield in the charge process.

CES is a coupled thermomechanical-based energy storage technology, which is likely to be suitable for applications with tens to hundreds megawatt power and tens megawatt-hour to a few gigawatt-hour capacity.

The benefits of the use of CES technology include time-shift of electric energy, reduction of the needs for new-generation installations, load following, increase of renewable penetration, and location shift of renewable generation [19].

8.1.3 Experimental Results of the World First CES Pilot Plant

As mentioned before, the world's first CES pilot plant has a rated power of 350 kW and a rated capacity of 2.5 MWh; see Table 8.1 for a summary of main parameters and Fig. 8.3 for photos of the pilot plant. The pilot plant is equipped with a cold thermal store (CTS) and a hot-water-based thermal energy store (TES). The CTS consists of packed bed of rocks (quartzite) with an average size of 15 mm for recovering the cold energy of liquid air at the cryogenic temperature range during the discharge process. The TES is for the provision of heat for heating liquid air during the discharge process with the heat supplied from the university combined heat and power plant because the pilot plant does not have the compression heat recovery function.

A number of runs have been carried out to study the performance of CES pilot plant [20]. Fig. 8.4 presents an example of the plant power output during a test. One can see that the system was initially at an idle condition, namely, no power output, before a sudden ramp-up request for power from the grid. Fig. 8.5 illustrates a detailed experimental run with a power ramp during discharge. A fast ramp-up of the plant can be seen to deliver 250 kW, ~70% of the maximum load in ~3 min, suggesting that the CES technology could be an option for fast reserve market.

Table 8.1 Main Parameters of the World's First CES Pilot Plant

Parameter	Value	Notes
Rated power	350 kW	With the integration of waste heat at 60°C
Rated storage capacity	2.5 MWh	Tank with a capacity of 60 m^3
Max liquid air production rate	1.44 t/h	
Max compression pressure	12 bar	
Max pressure of turbine inlet	60 bar	
Specific liquefaction work	0.73 MWh/t	Without cold recycle

Fig. 8.3 The CES pilot plant at the University of Birmingham (left photo shows the full plant picture; right photo shows the plant in operation).

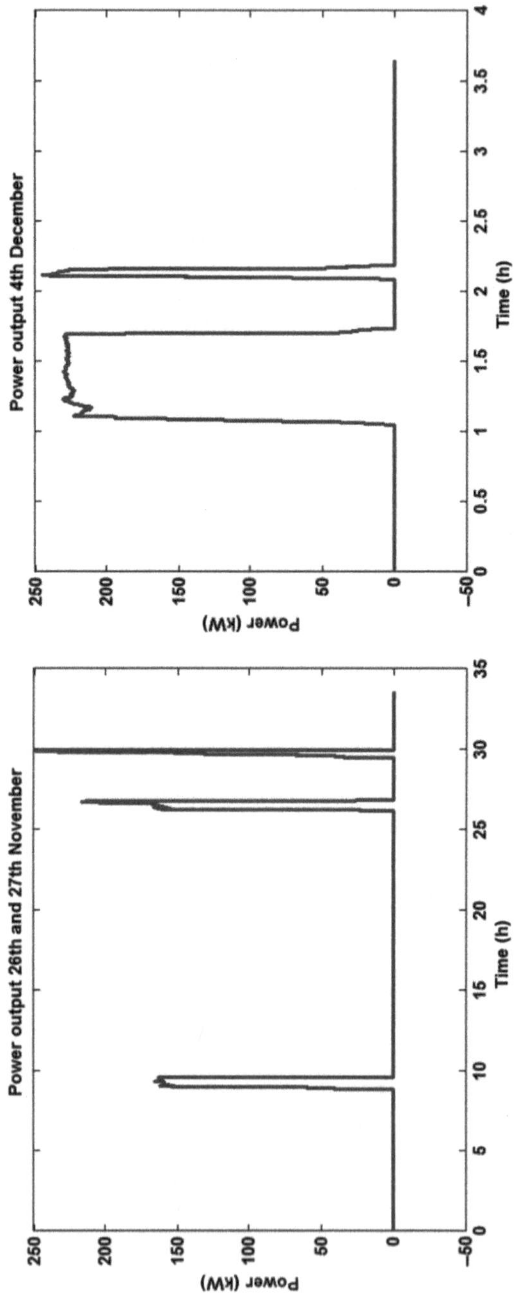

Fig. 8.4 CES pilot plant power output during discharge trials on different dates.

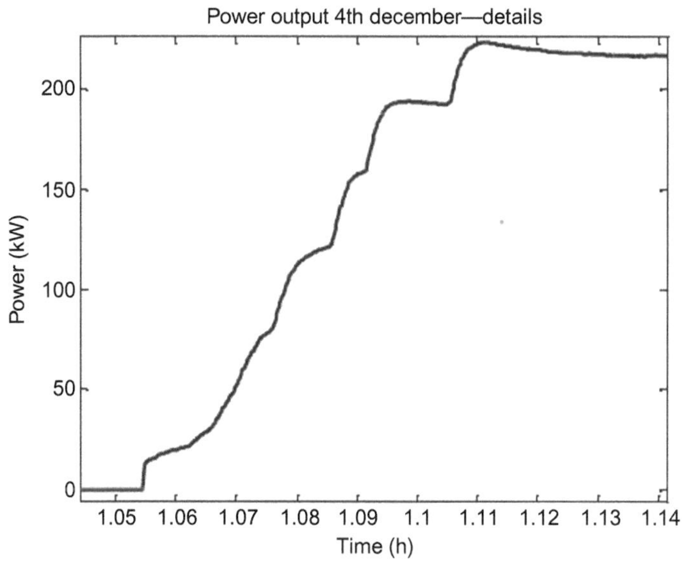

Fig. 8.5 CES pilot plant power output ramp.

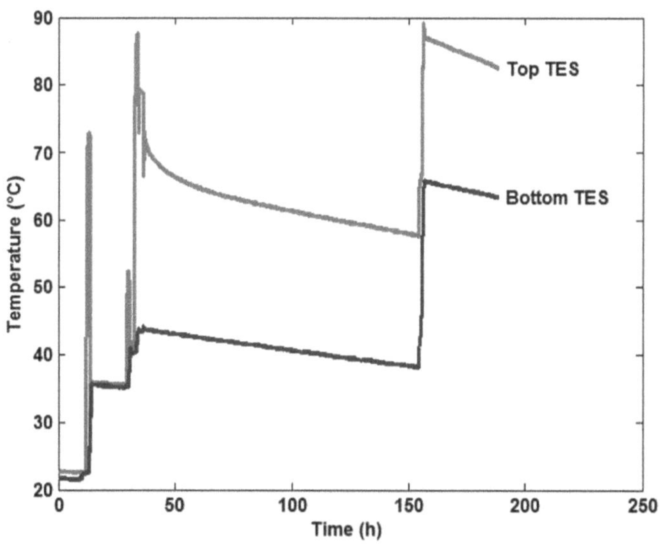

Fig. 8.6 Time evolution of temperature in the hot water TES.

Fig. 8.6 shows the temperature profiles of the top and bottom parts of the hot-water-based TES. The unit was held in an idle condition for about 100 h with an aim to assess the self-discharge of TES. A temperature drop of about 0.1°C/h is seen at both the top and bottom parts suggesting a small discharge

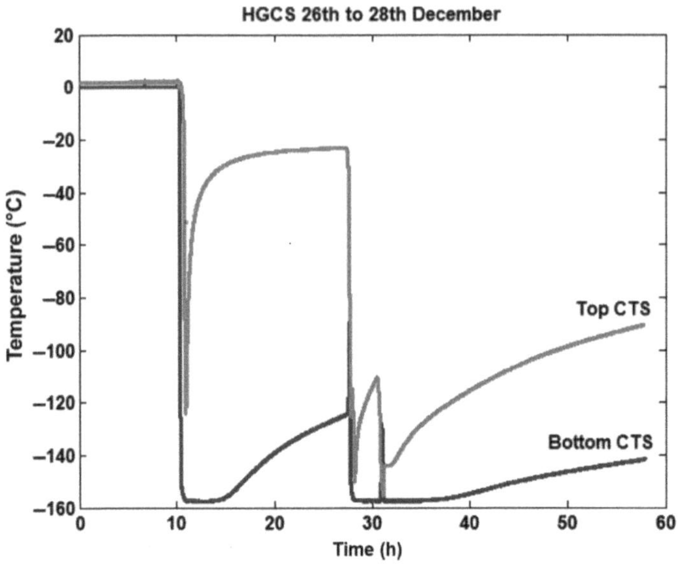

Fig. 8.7 Time evolution of temperature in the CTS.

rate. The temperature profiles also show a clear temperature stratification. The temperature profiles of the cold store of a test run are shown in Fig. 8.7. Clearly, the cold store was not fully charged during this run due to short discharge time. The data also suggest that the cold store can capture and store high-grade cold at a temperature close to −160°C.

8.1.4 Comparison of CES With Other Major Large Scale Energy Storage Technologies

Table 8.2 compares the major technical features between the CES and two other large-scale energy storage technologies—compressed air energy storage (CAES) and pumped hydro storage (PHS). The CES technology has a clear advantage of much higher volumetric and mass energy storage densities compared with CAES and PHS. Such a feature of CES makes it possible to store large amount of energy in large artificial tanks at a cost-effective manner. Due to the removal of the geologic and geographic constraints, the CES technology is highly flexible. The CES technology has a storage duration of hours to months and a response time of a few minutes, similar to the CAES and PHS technologies [21,25]. Another specific feature of the CES technology is its efficiency of utilizing waste-heat resources. As the air/nitrogen has a relatively low critical point than steam, waste heat from, for example, the exhaust of a simple-cycle gas turbine could be turned into electric energy

Table 8.2 Comparison of Technical Features of Major Large-Scale Energy Storage Technologies [21]

Storage Technology	PHS	CAES	CES
Energy density (Wh/L)	0.5–1.5, 1–2 [22]	3–6, 2–6 [22]	120–200
Specific energy (Wh/kg)	0.5–1.5	30–60	150–250
Power rating (MW)	100–5000	5–300	0.1–300
Daily self-discharge (%)	Very small	Small	0.5%–1.0%
Lifetime (years)	40–60	20–40	25+ [23]
Round-trip efficiency (%)	70–85	68–79	40–60, 55–80+ [23]
Suitable storage duration	Hours to months	Hours to months	Hours to months
Response time	Minutes [24]	Minutes [24]	Minutes [25]
Discharge time at power rating	1–24h+	1–24h+	1–8h
Power capital cost ($/kW)	600–2000	400–800, 800–1000 [26]	200–300, 900–1900 [23]
Energy capital cost ($/kWh)	5–100	2–50, 2–120 [27]	3–30, 260–530 [23]
Geologic constraints	Strict	Strict	None
Maturity	Mature	CAES commercialized Adiabatic CAES under development	Commercial demonstration

at an efficiency ~4 times that could be achieved by an organic Rankine cycle. This feature makes the integration of the CES technology with nuclear power plant (NPP) a highly competitive option, which could not be achieved by any other storage technologies. A disadvantage of the CES technology lies in its relatively low efficiency, up to ~60% using today's technology if the system is stand-alone. However, recent developments in the area by the authors' team at the University of Birmingham show that the efficiency can be enhanced by ~9%–12% by more efficient use of the compression heat.

8.2 INTEGRATION OF CES WITH NPPs

8.2.1 The Drive for Integration of CES With NPPs

NPPs are featured by high capital and low operating costs. This implies that the costs of electricity from such a capital-intensive technology can be low if operated at full capacity, and as a result, NPPs have been mainly used for baseload generation. However, with an increasing number of installations of NPPs, the capacity of power generation may exceed the baseload of power grids. For example, nuclear power contributes to ~53% of French total installation capacity while generating ~79% of the country's electricity [28,29]. The excess electricity at off-peak times has to be either exported to other countries or stored for late use (time-shift). If the two measures fail to balance the generation and demand, the NPPs have to be downregulated. When the NPPs operate at a part load, the cost of electricity production becomes high. Furthermore, frequent changes in the load can lead to quick aging of the equipment and affect the plant performance, leading to both economic- and safety-related problems [30].

Currently, pressurized water NPPs account for a large portion of the world's NPPs [31,32]. Downregulation of the load of the pressurized water NPPs in cases of low demands includes the control assemblies being inserted into the reactor vessel and associated changes of the coolants [33]. Apart from the safety issue and lifetime reduction, this operation mode also suffers two challenges in load following. First is the limited power changing gradient, which normally takes a few hours to achieve about half load. Second is that the downregulation of NPPs only balances the generation and demand at trough hours while other plants such as gas-fired power stations have to be employed to meet the peak demands.

8.2.2 The Principle and Flow Diagram of the Integration of CES With NPPs

A considerable amount of efforts has been made to deal with the load shift of NPPs, and the conventional method is the use of pumped hydro energy storage. Recent decades have seen the development of new approaches to the use of excess electricity for maintaining the NPPs at nearly the full load. These include steam accumulator storage [34], large-scale hydrogen production and storage [35,36], and geothermal heat storage [37]. The integration of the CES with NPPs has been proposed recently by the authors of this chapter, which has the potential to resolve the issues associated with the load regulation of NPPs in a more economical and efficient manner. Fig. 8.8 shows the principle of the integrated system, which consists of an NPP subsystem and the CES subsystem [38]. The NPP subsystem in the integrated system is similar to the conventional pressurized water NPP. The only difference lies in that there are two three-way valves in the secondary loop, which enables the working fluid to feed into either the steam turbine to produce electricity or heat exchanger 4 to superheat high-pressure air in CES subsystem (Fig. 8.8).

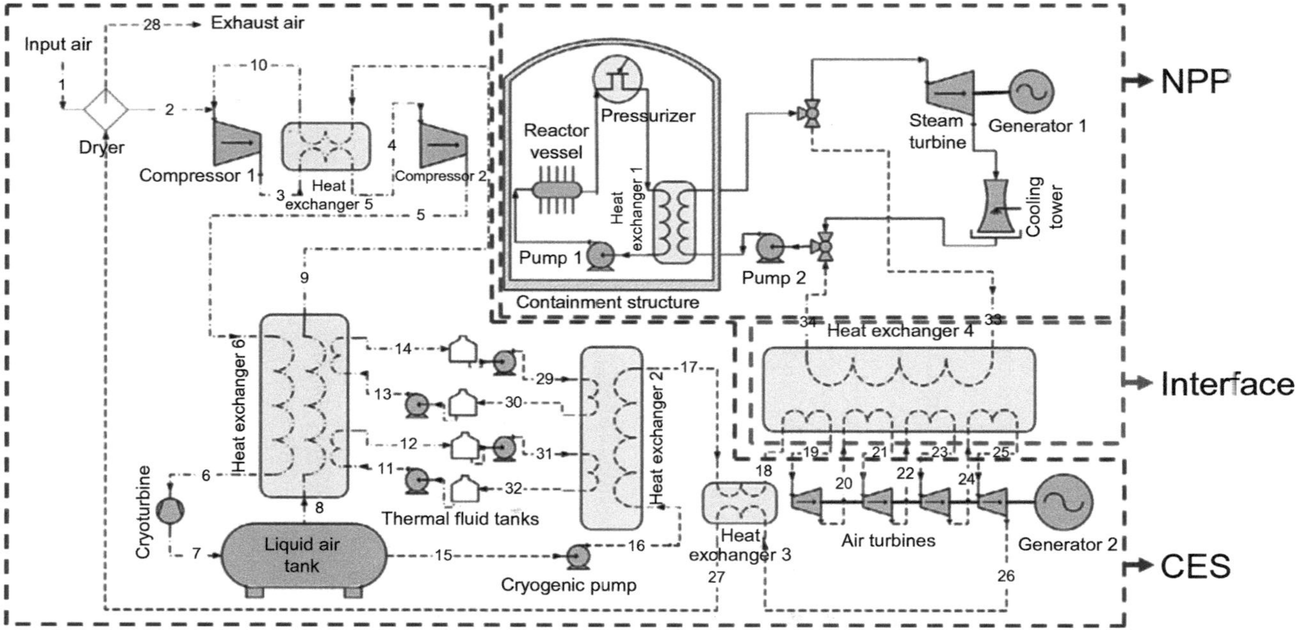

Fig. 8.8 An integrated NPP-CES technology, illustrating the linkage between the NPPs and CES subsystems via heat exchange 4.

The CES subsystem consists of an air liquefaction unit in the left part and a power recovery unit in the right-bottom part of Fig. 8.8. The air liquefaction subsystem works in a similar way as the simplest Linde-Hampson liquefier except for the use of external cold energy through heat exchanger 6. The cold storage and associated recovery function acts as a bridge between the air liquefaction unit and the power recovery unit. Propane and methanol are taken as cold storage media in the assessment, which also act as working fluids for heat transfer. The rationales for considering the two cold storage media are because they meet well the required temperature range and have a high-specific heat capacity. A two-tank configuration is proposed for storing and recovering the cold energy for each of the two fluids: During cold storage step, the two fluids are pumped respectively from their warm tanks to their cold tanks (cold energy is stored, while power is recovered); during the cold recovery step, the two fluids flow respectively from their cold tanks to their warm tanks (cold energy is released, while power is stored). Additional benefits for the use of the cold storage fluids for both transferring and storing cold energy include a greater simplification of the system design with no additional heat exchangers and more straightforward operating strategy with the amount of cold energy and the objective temperature easily regulated by controlling the flow rate of the fluids, which are extremely difficult to achieve using the conventional packed pebble bed-based method.

The power recovery unit is coupled with the NPP through heat exchanger 4 where low-grade heat from the NPP is utilized. The use of such an approach enables the thermal energy, normally wasted in the cooling process, to be converted to electric energy at a high efficiency, which cannot be achieved by any other storage technologies.

The heat exchanger 4 provides a linkage between the CES subsystem and the NPP subsystem. Such an integration allows the reactor core and the primary loop of NPPs to operate steadily at full load at all times, while the net output power is adjusted only by the CES subsystem. As the power recovery unit in the CES subsystem is similar to a power generation process using a gas turbine, a much faster rate of power change could be achieved in comparison with the conventional regulation of NPPs.

8.2.3 The Operation Modes of the Integrated CES-NPP System

The integrated CES-NPP system could have three operational modes depending on the supply-demand balance: electric energy storage mode, electric energy recovery mode, and conventional mode. These modes are described in the following.

8.2.3.1 Electrical Energy Storage Mode

This mode occurs at trough hours when the demand is much lower than the rated power of the NPP. In this mode, the NPP operates in a traditional way to drive the steam turbine to produce electricity, and the excessive power is used to drive the air liquefaction unit to produce liquid air (electric energy is stored in liquid air). In this process, dry airstream (2) and return gas stream (10) are mixed and compressed to an elevated pressure (5) by a two-stage compressor (compressor 1 and compressor 2) with intercooling (3–4) in heat exchanger 5. After rejecting heat via heat exchanger 6 in the main cold box (processes 5–6), the high-pressure air is cooled to the lowest temperature level, followed by a near-isentropic expansion process in the so-called cryoturbine to give liquid air. A fraction of the product is vaporized in the cryogenic tank and introduced back to the main cold box through heat exchanger 6 and the intercooler heat exchanger 5 to supply part of the cold energy. The remaining cold requirement of the main cold box is met by the cold stored in the cold storage tanks, namely, the cold energy is discharged when electric energy is charged.

8.2.3.2 Electrical Energy Release Mode

This mode of operation occurs at peak hours when the demand is higher than NPP's rated power. The recovery unit is turned on to produce additional power. In this process, liquid air in liquid air tank is pressurized first by the cryogenic pump. The high-pressure air transfers its cold energy to the cold storage media in processes 16–17 via heat exchange 2 followed by a preheating process in heat exchanger 3 by the exhaust gas stream (processes 26–27). In the meantime, the secondary coolant in the NPP is introduced into heat exchanger 4 by switching on the three-way valves. This leads to a further increase in the temperature and pressure of the air, which, upon expansion in a four-stage air turbine with interheating in heat exchanger 4, drives generator 2 to produce electricity. The exhaust air is used first to preheat the high-pressure air in heat exchanger 3 and then to regenerate the desiccant in dryer.

8.2.3.3 Conventional Operation Mode

This mode of operation occurs at nontrough and nonpeak hours. In this mode, both the air liquefaction unit and the power recovery unit are switched off so that the NPP operates in a conventional way to drive the steam turbine to produce electricity. With thermal insulation technologies, cold dissipation from liquid air tank and cold storage tanks is expected to be low.

8.2.4 Performance Assessment of the Integrated CES-NPP System

8.2.4.1 Performance of the Integrated CES-NPP System

The integrated system is evaluated by model simulations using an in-house developed package called Thermal System Optimal Designer (TSOD) under the MATLAB environment. For simplification and illustration of the concept, the integrated system is assumed to run at the steady state in each of the operating modes with negligible pressure drop and heat loss during the heat transfer process. Thermal properties used in the TSOD are calculated using a commercial software, REFPROP 8.0, developed by the National Institute of Standards and Technology (NIST). All components are modeled by using energy and mass balances with the pinch constraints in multiflow heat exchangers, namely, the second law of thermodynamics, being respected.

One of the most important performance indicators of an energy storage technology is its round-trip efficiency, which has already been defined in Section 8.1.2 (see Eq. 8.1). Application of Eq. (8.1) to the integrated CES-NPP system gives

$$\eta_{RT} = \frac{(W_{ER} - W_C) \cdot t_{ER}}{W_{ES} \cdot t_{ES}} \tag{8.5}$$

where W_{ER} and W_C are, respectively, the power output in the electric energy release mode and the conventional operation mode; W_{ES} is the power consumption for liquid air production in the electric energy storage mode; and t_{ER} and t_{ES} correspond to the electric energy release and storage durations. This round-trip efficiency is selected as an objective function in the parametric optimization.

Table 8.3 summarizes all the assumptions used for the performance assessment together with the component performance data. Note that, for clarity and simplification without losing the most important features of the system, power consumption for pumping the cold storage fluids and driving the air dryer has not been included as they are much lower compared with other components.

Taking a small-scale NPP rated at 250 MW as an example, the parametric optimization gives the flow rate, temperature, and pressure data as listed in Table 8.4 (refer to Fig. 8.8 for the flow numbers). The air liquefaction capacity for such an integrated system is 150 kg/s, which is equivalent to 540 t/h. This scale of liquefaction operation has been used commercially for liquefied natural gas (LNG) [39].

Table 8.3 List of Assumed Operating Conditions and Component Performance Data

Rated power of the NPP W_C (MW)	250
Secondary loop topping pressure of the NPP P_{NPP} (kPa)	7093
Thermal efficiency of the NPP η_{NPP} (%)	31
Ambient pressure P_a (kPa)	101
Ambient temperature P_a (kPa)	288
Liquid air storage pressure P_a (kPa)	101
Operational period in energy storage mode t_{ES} (h/day)	8
Operational period in energy release mode t_{ER} (h/day)	1
Operational period in conventional mode t_C (h/day)	15
Temperature approach of heat exchangers (K)	2
Isentropic efficiency of the air turbines η_{AT} (%)	92
Isentropic efficiency of the cryoturbine η_{CT} (%)	88
Isothermal efficiency of air compressors η_{AC} (%)	90
Isentropic efficiency of the cryogenic pump η_{CP} (%)	70

Table 8.4 The Mainstream Parameters of the NPP-CES Integrated System

Flow No.	Mass Flow (kg/s)	Pressure (kPa)	Temperature (K)	Fluid
1	150	101	288	Air
2	150	101	288	Air
3	179	1159	288	Air
4	179	1159	282	Air
5	179	13,409	288	Air
6	179	13,409	102	Air
7	179	101	81	Air
8	29	101	83	Air
9	29	101	250	Air
10	29	101	288	Air
11	167	101	95	Propane
12	167	101	212	Propane
13	90	101	219	Methanol
14	90	101	286	Methanol
15	1195	101	80	Air
16	1195	11,385	83	Air
17	1195	11,385	283	Air
18	1195	11,385	380	Air
19	1195	11,385	553	Air
20	1195	3497	396	Air
21	1195	3497	553	Air
22	1195	1074	396	Air
23	1195	1074	553	Air
24	1195	330	397	Air

Continued

Table 8.4 The Mainstream Parameters of the NPP-CES Integrated System—cont'd

Flow No.	Mass Flow (kg/s)	Pressure (kPa)	Temperature (K)	Fluid
25	1195	330	553	Air
26	1195	101	397	Air
27	1195	101	288	Air
28	1195	101	288	Air
29	723	101	288	Propane
30	723	101	217	Propane
31	1337	101	214	Methanol
32	1337	101	93	Methanol
33	442	7093	560	Steam
34	442	7093	493	Steam

Compared with large-scale LNG production, air liquefaction in the proposed system is simpler as most of the cold energy is supplied by the cold storage unit instead of external cold production in LNG production using cascade, mixed-refrigerant, or N_2 expander cycle methods. Furthermore, cold storage and recovery using thermal fluids enable very efficient heat transfer process and as a result significantly decrease the power consumption in the air liquefaction process. Table 8.4 also shows that the high-pressure feed air is cooled to 102 K before the expansion in the cryoturbine where ~84% of the feed air is liquefied (compared with the Linde-Hampson liquefier that is only ~7% at the same topping pressure). The main feed air cooling is a supercritical process at about 13,409 kPa (the critical pressure of air is ~3770 kPa), while the topping pressure in power recovery process is slightly lower at ~11,385 kPa.

Tables 8.5 and 8.6 show the results of the analyses of the integrated system including power/heat and exergy for key components. The exergy loss ratio in the table is defined as the percentage of exergy loss to the overall power/exergy changes. The net power consumption in the electric energy storage mode is 76.74 MW, while about two-thirds of the exergy loss occur in the power transfer processes (air compression). The exergy loss in the main heat transfer process in heat exchanger 6 contributes less than one-third with a low exergy loss ratio of about 8%. Although the exergy loss ratio in heat exchanger 5 is high, the total exergy loss in this process is only ~0.07 MW as the grade of thermal energy is low at close to the ambient temperature.

In the electric energy release mode, the energy changes occur mainly in heat transfer processes for both the cold recovery in heat exchanger 2 and

Table 8.5 Performance of Key Components.

Components	Power/Heat Loads (MW)	Exergy Loss (MW)	Exergy Loss Ratio (%)
Energy storage mode			
Compressor 1	40.03	3.96	10
Heat exchanger 5	1.11	0.07	87
Compressor 2	39.83	3.99	10
Heat exchanger 6	59.71	3.74	8
Cryoturbine	3.12	0.43	12
Energy release mode			
Cryogenic pump	19.18	5.77	30
Heat exchanger 2	439.79	38.76	10
Heat exchanger 3	133.02	2.012	10
Heat exchanger 4	807.91	78.45	20
Air turbines	706.69	62.81	8

Table 8.6 Performance of the Overall System.

Net power consumption in energy storage mode (MW)	76.74
Net power output in energy release mode (MW)	687.51
Round-trip efficiency (%)	71.26

superheating of the high-pressure air in heat exchanger 4. The exergy loss ratio in heat exchanger 4 is approximately 20% mainly due to phase change, leading to the pinch point constraint and hence lowering the efficiency. The exergy loss in cold recovery process in heat exchanger 2 is low (~10%) due to the use of thermal fluids. The net power output in the electric energy release mode is 687.51 MW, corresponding to a round-trip efficiency of ~71%. Based on this operation strategy, the NPP-CES integrated system could generate 173.26 MW electric power at the trough hours (8 h/day), 687.51 MW at peak hours (1 h/day), and 250 MW at the remaining hours.

It should be noted that the electric energy release unit in the integrated system could operate without the NPP unit while being heated by ambient or other heat sources. Such an operation can extend the operation hours by adjusting the mass flow rate in the electric energy release mode although the net power output decreases significantly and there is a decrease in the round-trip efficiency. However, these results also reveal that the electric energy release unit can also be used as the backup unit in the NPPs (instead of expensive and short-life-span battery and diesel-genset-based backup units).

8.2.4.2 Performance Enhancement Through Adjusting Operating Conditions

There are a number of ways to enhance the efficiency of the integrated CES-NPP system including changes to the system configuration and changes to the operating conditions. In the following, discussion is made on how the operating conditions affect the system performance.

First, to look at the storage pressure effect. Thermodynamics tells that the higher the pressure, the easier the gas can be liquefied, and hence, a higher CES round-trip efficiency could be achieved [14]. The influence of the storage pressure is therefore investigated on the performance of the integrated NPP-CES system. Fig. 8.9 shows the results. An increase in the storage pressure shows a linear increase in the round-trip efficiency, and the increase rate is about 0.8% per bar increased, indicating that the storage pressure affects mainly the liquefaction process as both the net output power and the mass flow rate of liquid air in the electric energy release mode change only slightly with increasing storage pressure. Here, the almost constant net output power is because the increased storage pressure reduces only the power consumption of the cryogenic pump.

Second, to examine the effect of the secondary loop topping pressure of the NPP in electric energy release mode. The steam in the secondary loop of the NPP is used to superheat the high-pressure air, instead of driving the

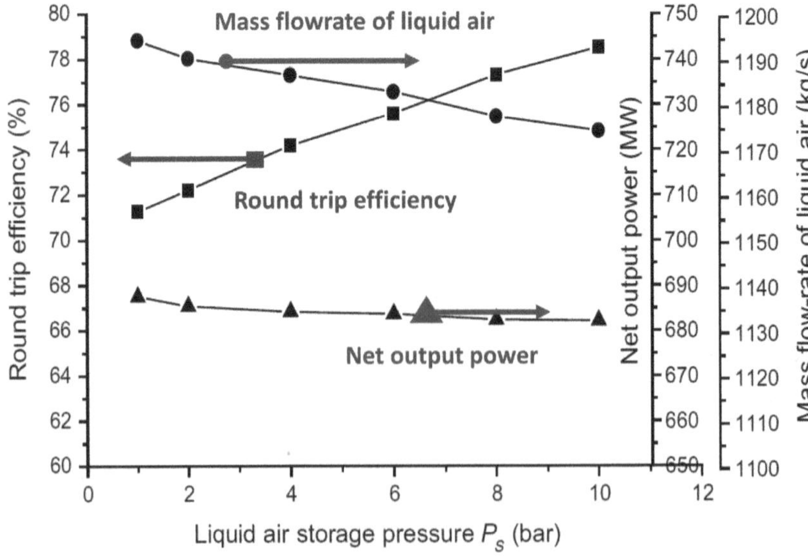

Fig. 8.9 The influence of liquid air storage pressure on the system performance.

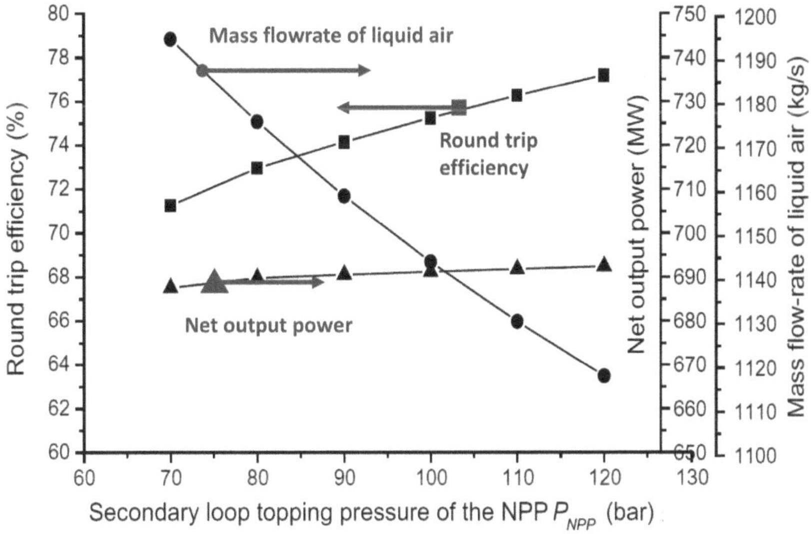

Fig. 8.10 The influence of secondary loop topping pressure of the NPP on the system performance.

steam turbine, in the electric energy release mode. Fig. 8.10 shows the results, which demonstrates clearly an almost linear increase in the round-trip efficiency with increasing secondary loop topping pressure. Different to the influence of the liquid air storage pressure, the mass flow rate of liquid air in the electric energy release mode decreases with increasing secondary loop topping pressure. This indicates that the secondary loop topping pressure affects mainly on the electric energy release mode. Although the mass flow rate of liquid air decreases, the net output power is almost independent of the topping pressure.

It should be noted that the increase in either the liquid air storage pressure or the NPP secondary loop topping pressure could face technical and economic challenges. For example, currently, cryogenic tanks are typically operated at a pressure below ~10 bar at a ~100 t scale and generally at the atmospheric pressure for over 100 t scales. As a consequence, the selection of the working pressures should consider the balance between the system performance and the capital, operation, and maintenance costs.

Third, to investigate the ambient temperature on the system performance, Fig. 8.11 shows the results. The round-trip efficiency decreases by ~2% for an every 5°C increase in the ambient temperature. Similar to the effect of liquid air storage pressure, the ambient temperature has a much stronger effect on the air liquefaction process than that on the electric energy

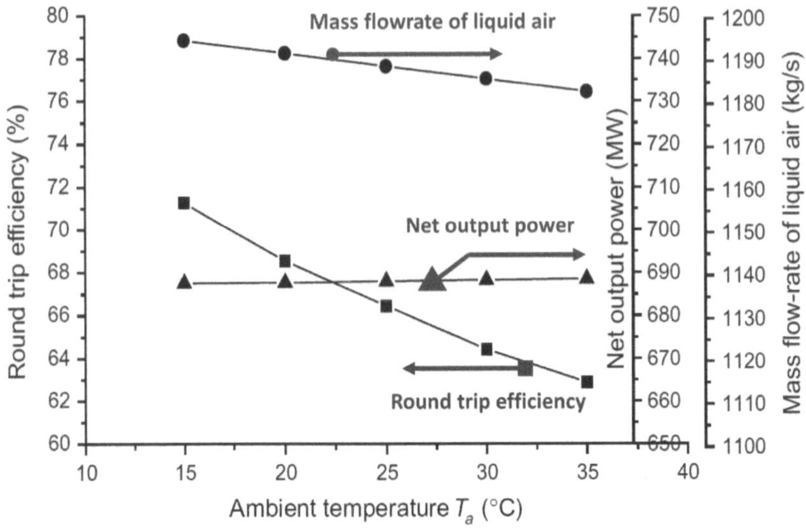

Fig. 8.11 The influence of the ambient temperature on the system performance.

release process in terms of the net output power. The mass flow rate of liquid air in the electric energy release mode changes only slightly with increasing ambient temperature. These results also indicate that the locations with a low ambient temperature are preferred for the installation of NPP-CES integrated systems.

The above analyses also indicate that the net output power in the electric energy release mode changes only slightly at ~690 MW. As a result, the integrated system is capable of generating some 2.7 times the power at peak demands compared with an NPP system alone, regardless of the location of the installation and the operating pressures.

8.3 SUMMARY OF THE CHAPTER

The CES technology has now been progressed to the commercial demonstration stage since invented by some of the authors of this chapter 12 years ago. This technology is characterized by low capital and maintenance costs, high-energy density, long life span, and being site-constraint-free, hence providing a highly promising alternative to the large-scale site-dependent pumped hydro and compressed air energy storage technologies. The CES technology, upon integrated with NPPs, provides an effective and efficient solution to the load shift of the plants.

Thermodynamic analyses on the integrated CES-NPP system under fairly general baseline assumptions show that the round-trip efficiency of electric energy storage is about 71% while the net output power in the electric energy release mode could be as high as 2.7 times the NPP rated power, illustrating the potential of using the method to change the net power output (instead of downregulating) of the NPPs. Sensitivity analyses show that the storage pressure and ambient temperature mainly affect the air liquefaction process while the secondary loop topping pressure of the NPPs affects more on the energy release process.

REFERENCES

[1] Chen H, Ding Y, Peters T, Berger F. Energy storage and generation, US Patent US20090282840, 2009.

[2] E.M. Smith, Storage of electrical energy using supercritical liquid air, Proc. Inst. Mech. Eng. 191 (1977) 289–298.

[3] K. Kenji, H. Keiichi, A. Takahisa, Development of generator of liquid air storage energy system, Tech. Rev. Mitsubishi Heavy Ind. Ltd 35 (1998) 117–120.

[4] K. Chino, H. Araki, Evaluation of energy storage method using liquid air, Heat Transf. Asian Res. 29 (2000) 347–357.

[5] H. Araki, M. Nakabaru, K. Chino, Simulation of heat transfer in the cool storage unit of a liquid–air energy storage system, Heat Transf. Asian Res. 31 (2002) 284–296.

[6] Wakana H, Chino K, Yokomizo O. Cold heat reused air liquefaction/vaporization and storage gas turbine electric power system, US Patent US20030101728, 2005.

[7] Vandor D. System and method for liquid air production, power storage and power release, US Patent US8020404B2, 2011.

[8] R. Morgan, S. Nelmes, E. Gibson, G. Brett, Liquid air energy storage—analysis and first results from a pilot scale demonstration plant, Appl. Energy 137 (2015) 845–853.

[9] Y. Li, Y. Jin, H. Chen, C. Tan, Y. Ding, An integrated system for thermal power generation, electrical energy storage and CO2 capture, Int. J. Energy Res. 35 (2011) 1158–1167.

[10] H. Chen, Y. Ding, Y. Li, X. Zhang, C. Tan, Air fuelled zero emission road transportation: a comparative study, Appl. Energy 88 (2011) 337–342.

[11] Y. Li, X. Wang, Y. Jin, Y. Ding, An integrated solar-cryogen hybrid power system, Renew. Energy 37 (2012) 76–81.

[12] Y. Li, X. Wang, Y. Ding, A cryogen-based peak-shaving technology: systematic approach and techno-economic analysis, Int. J. Energy Res. 37 (2013) 547–557.

[13] G.L. Guizzi, M. Manno, L.M. Tolomei, R.M. Vitali, Thermodynamic analysis of a liquid air energy storage system, Energy 93 (2015) 1639–1647.

[14] A. Sciacovelli, A. Vecchi, Y.L. Ding, Liquid air energy storage (LAES) with packed bed cold thermal storage—from component to system level performance through dynamic modelling, Appl. Energy 190 (2017) 84–98.

[15] X. She, X. Peng, B. Nie, G. Leng, X. Zhang, L. Weng, et al., Enhancement of round trip efficiency of liquid air energy storage through effective utilization of heat of compression, Appl. Energy 206 (2017) 1632–1642.

[16] H. Peng, X. Shan, Y. Yang, X. Ling, A study on performance of a liquid air energy storage system with packed bed units, Appl. Energy 211 (2018) 126–135.

[17] Ding Y, Wen D, Dearman PT. Cryogenic engines, US Patent US20090320476, 2009.

[18] GE storage agreement with highview, Modern Power Systems (2014). April, http://viewer.zmags.com/publication/f997bf2b#/f997bf2b/4.

[19] Y. Li, H. Chen, X. Zhang, C. Tan, Y. Ding, Renewable energy carriers: hydrogen or liquid air/nitrogen? Appl. Therm. Eng. 30 (2010) 1985–1990.

[20] A. Sciacovelli, D. Smith, H. Navarro, Y. Li, Y. Ding, in: Liquid air energy storage—operation and performance of the first pilot plant in the world, The 29th International Conference on Efficiency, Cost, Optimization, Simulation and Enviromental Impact of Energy Systems Portoroz, Slovenia, 2016.

[21] H. Chen, T.N. Cong, W. Yang, C. Tan, Y. Li, Y. Ding, Progress in electrical energy storage system: a critical review, Prog. Nat. Sci. 19 (2009) 291–312.

[22] Anon, Electrical energy storage: white paper, Technical report. Prepared by electrical energy storage project team. International Electrotechnical Commission (IEC), 2011. December, http://www.iec.ch/whitepaper/pdf/iecWP-energystorage-LR-en.pdf.

[23] Highview Power Storage: Secure, Clean Power. Highview Power Storage, http://www.imeche.org/docs/default-source/2011-press-releases/Highview_2pager.pdf?sfvrsn=0, 2012.

[24] S.M. Shoenung, Characteristics and technologies for long- vs. short-term energy storage: a study by the DOE energy storage systems program, Sandia National Laboratories. United States Department of Energy, 2001. March, http://prod.sandia.gov/techlib/access-control.cgi/2001/010765.pdf.

[25] M. Akhurst, A. Atkins, I. Arbon, M. Ayres, N. Brandon, R. Bruges, et al., Liquid air in the energy and transport systems: opportunities for industry and innovation in the UK, Full report, The Centre for Low Carbon Futures, 9th May, http://liquidair.org.uk/files/full-report.pdf, 2013.

[26] Electricity storage: technology brief. Energy technology systems analysis programme, International Renewable Energy Agency, 2012. April, https://www.scribd.com/document/355792813/Electricity-Storage-Technology-Brief.

[27] P. Taylor, R. Bolton, D. Stone, X.P. Zhang, C. Martin, P. Upham, Pathways for energy storage in the UK, The Centre for Low Carbon Futures, 2012. 27th March, http://oro.open.ac.uk/40087/2/Pathways_for_Energy_Storage_in_the_UK.pdf.

[28] C. Bruynooghe, A. Eriksson, G. Fulli, Load-following operating mode at nuclear power plants (NPPs) and incidence on operatio and maintenance (O&M) costs. Compatibility with wind power variability, European Commission, Joint Research Centre, Institute of Energy, 2010.

[29] C. Coombs, French nuclear power: a model for the world? Hinckley J. Polit. 11 (2010) 7–13.

[30] A. Lokhov, Technical and economic aspect of load following with nuclear power plant, Nuclear Energy Agency, Organisation for Economic Co-operation and Development, 2011.

[31] E.E. Michaelides, Nuclear Power Plants: Alternative Energy Sources, Springer, Berlin, Heidelberg, 2012, pp. 131–172.

[32] M.V. Kothare, B. Mettler, M. Morari, P. Bendotti, C.M. Falinower, Level control in the steam generator of a nuclear power plant, IEEE Trans. Control Syst. Technol. 8 (2000) 55–69.

[33] H. Ludwig, T. Salnikova, A. Stock, Load cycling capabilities of German nuclear power plants, Int. J. Nucl. Power 55 (2010) 2–8.

[34] P.V. Gilli, G. Beckman, Steam storage adds peaking capacity to nuclear plants, Energy Int. 10 (1973) 16–18.

[35] B. Yildiz, M.S. Kazimi, Efficiency of hydrogen production systems using alternative nuclear energy technologies, Int. J. Hydrog. Energy 31 (2006) 77–92.

[36] C.W. Forsberg, M.S. Kazimi, Nuclear Hydrogen Using High-Temperature Electrolysis and Light Water Reactors for Peak Electricity Production, Massachusetts Institute of Technology, 2009.
[37] C.W. Forsberg, Y. Lee, M. Kulhanek, M.J. Driscoll, in: Gigawatt-year nuclear-geothermal energy storage for light-water and high-temperature reactors, International Congress on the Advances in Nuclear Power Plants, Chicago, IL, 2012.
[38] Y. Li, H. Cao, S. Wang, Y. Jin, D. Li, X. Wang, et al., Load shifting of nuclear power plants using cryogenic energy storage technology, Appl. Energy 113 (2014) 1710–1716.
[39] W. Cao, X. Lu, W. Lin, A. Gu, Parameter comparison of two small-scale natural gas liquefaction processes in skid-mounted packages, Appl. Therm. Eng. 26 (2006) 898–904.

Index

Note: Page numbers followed by *f* indicate figures and *t* indicate tables.

F

G

O

P

S

T

U

FSC
www.fsc.org
MIX
Papier | Fördert
gute Waldnutzung
FSC® C083411

Kolibri 360 GmbH
Ferdinand-Jühlke-Straße 7
99095 Erfurt, Deutschland
produktsicherheit@kolibri360.de